# The Tropospheric Transport of Pollutants and Other Substances to the Oceans

Prepared by the
Workshop on Tropospheric Transport of
     Pollutants to the Ocean Steering Committee
Ocean Sciences Board
Assembly of Mathematical and Physical Sciences
National Research Council

NATIONAL ACADEMY OF SCIENCES
Washington, D.C.  1978

**Library of Congress Cataloging in Publication Data**

NRC Workshop on Tropospheric Transport of Pollutants to the Ocean Steering Committee. 1975.

The tropospheric transport of pollutants and other substances to the ocean.

Bibliography: p.
1. Marine pollution. 2. Troposphere. 3. Air—Pollution. I. National Research Council. Assembly of Mathematical and Physical Sciences. II. Title.

GC1085.N24 1975        628.1′686′162        78-2789
ISBN 0-309-02735-7

*Available from:*

Printing and Publishing Office
National Academy of Sciences
2101 Constitution Avenue, N.W.
Washington, D.C. 20418

Printed in the United States of America

# Workshop Staff

*Steering Committee*
JOSEPH M. PROSPERO, *Chairman*
EDWIN F. DANIELSEN
ROBERT A. DUCE
EDWARD D. GOLDBERG

*Panel Chairmen*
EDWIN F. DANIELSEN and JAMES W. DEARDORFF, Modeling the Atmospheric Transport of Pollutants and Other Substances from Sources to the Oceans
JAMES P. FRIEND, Nonhydrocarbon Gases
WILLIAM D. GARRETT, Petroleum and Related Natural Hydrocarbons
REINHOLD A. RASMUSSEN, Halogenated Hydrocarbons
W. GEORGE N. SLINN, Wet and Dry Removal Processes
HERBERT L. VOLCHOK, Radionuclides
DOUGLAS M. WHELPDALE, Techniques
WILLIAM H. ZOLLER, Metals

*Ocean Sciences Board*
RICHARD C. VETTER, *Executive Secretary*
MARY LOU LINDQUIST, *Assistant Staff Officer*

iii

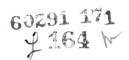

# Participants and Contributors

TERRY F. BIDLEMAN, University of South Carolina
ROGER CHESSELET, CFR/CNRS, Gif-sur-Yvette, France
ROY CHESTER, University of Liverpool, Liverpool, England
THOMAS CHURCH, National Science Foundation
ROBERT CITRON, The Smithsonian Institution
EDWIN F. DANIELSEN, National Center for Atmospheric Research, Oregon State University
DOUGLAS D. DAVIS, University of Maryland
JAMES W. DEARDORFF, National Center for Atmospheric Research
DENNIS DEAVEN, National Meteorological Center
ROBERT A. DUCE, University of Rhode Island
JAMES P. FRIEND, Drexel University
WILLIAM D. GARRETT, Naval Research Laboratory
EDWARD D. GOLDBERG, Scripps Institution of Oceanography, University of California
JURGEN H. HAHN, Max-Planck Institut für Chemie, Mainz, West Germany
LUTZ HASSE, Universität Hamburg, Hamburg, Germany
BRUCE B. HICKS, Argonne National Laboratory
GLENN R. HILST, The Research Corporation of New England
AUSTIN W. HOGAN, State University of New York, Albany
JACQUES JEDWAB, University of Brussels, Brussels, Belgium
C. S. KIANG, National Center for Atmospheric Research
DEVENDRA LAL, Physical Research Laboratory, Navrangpura, India

v

EUGENE E. LIKENS, Cornell University
MARY LOU LINDQUIST, National Research Council
PETER S. LISS, University of East Anglia, Norwich, England
LESTER MACHTA, Environmental Research Laboratories, National
  Oceanic and Atmospheric Administration
JERALD D. MAHLMAN, Geophysical Fluid Dynamics Laboratory,
  Princeton University
EDWARD A. MARTELL, National Center for Atmospheric Research
K. O. MUNNICH, Heidelberg University, Heidelberg, West Germany
DONALD H. PACK, Environmental Research Laboratories, National
  Oceanic and Atmospheric Administration
CLAIR C. PATTERSON, California Institute of Technology
DOUGLAS H. PEIRSON, Atomic Energy Research Establishment,
  Harwell, England
D. PIEROTTI, Washington State University
JOSEPH M. PROSPERO, University of Miami
REINHOLD A. RASMUSSEN, Oregon Graduate Center
ELMAR R. REITER, Colorado State University
ELMER ROBINSON, Washington State University
GEORGE A. SEHMEL, Battelle Memorial Institute, Richland
WOLFGANG SEILER, Max-Planck Institut für Chemie, Mainz, West
  Germany
W. GEORGE N. SLINN, Battelle Memorial Institute, Richland
JACK W. SWINNERTON, Naval Research Laboratory
SHIZUO TSUNOGAI, Hokkaido University, Hakodate, Japan
KARL K. TUREKIAN, Yale University
RICHARD C. VETTER, National Research Council
OTTAVIO VITTORI, Laboratorio Microfisica dell'Atmosfera, C.N.R.,
  Bologna, Italy
HERBERT L. VOLCHOK, Health and Safety Laboratory, Department of
  Energy
DOUGLAS M. WHELPDALE, Atmospheric Environment Service,
  Ontario, Canada
PETER E. WILKNISS, Naval Research Laboratory
JOHN W. WINCHESTER, Florida State University
WILLIAM H. ZOLLER, University of Maryland

# Preface

Marine scientists have become increasingly concerned over the possible effects of pollutants in the marine environment. Concern is aroused as much by the lack of knowledge about pollutants and their effects in, and over, the oceans as it is by specific known cases of deleterious impacts. Responding to these concerns, an Executive Meeting of the Scientific Committee on Oceanic Research (SCOR) of the International Council of Scientific Unions (ICSU), in December 1974, called for an international workshop to assess the problem. In recognition of the fact that much of the research is carried out by scientists in the United States, SCOR suggested that such a meeting be organized and held under the aegis of the Ocean Sciences Board (OSB) of the National Research Council, which serves as the U.S. National Committee to SCOR. The OSB agreed to sponsor the meeting. Support for the meeting and for participation by U.S. scientists was provided by the National Science Foundation. Participation by scientists from outside the United States was supported by the United Nations Environment Program through the Intergovernmental Oceanographic Commission.

In March 1975, an *ad hoc* Steering Committee was established (Edwin Danielsen, National Center for Atmospheric Research; Robert Duce, University of Rhode Island; Edward D. Goldberg, University of California; Joseph M. Prospero, University of Miami). After a few discussions, it was clear that there were two major subject areas: (1) specific pollutant and trace substance data and (2) transport and removal processes. The first area was clearly the domain of chemists and the second, of atmospheric scientists. It was decided to organize the Workshop participants into these two main categories and to set up panels within each. The panels and their chairmen in the first category, "Substances," were Metals (W. H. Zoller), Halogenated Hydrocarbons (R. A. Rasmussen), Petroleum Hydrocarbons (W. D. Garrett), Nonhydrocarbon Gases (J. Friend), and Radionuclides (H. Volchok). The second category, or "Processes," consisted of two panels: Modeling the Atmospheric Transport of Pollution and Other Substances from Sources to the Oceans (E. Danielsen and J. Deardorff, Co-Chairmen) and Wet and Dry Removal Processes (W. G. N. Slinn). Eventually, a third category consisting of one panel was established: Techniques (D. Whelpdale, Chairman).

In selecting participants, an effort was made to invite those individuals who had a broad view of the general problem as well as expertise in a specific area. Persons invited to the Workshop were required to prepare and submit, prior to the meeting, position papers in the specialty fields. Copies of these papers were bound in a volume that was made available to all participants and that served as a reference work for the panel discussions. (A limited number of copies of this volume is available through the Ocean Sciences Board, National Research Council, as *Background Papers for a Workshop on the Tropospheric Transport of Pollutants to the Ocean.*)

The Workshop was held on December 8–12, 1975, on the campus of the Rosenstiel School of Marine and Atmospheric Science, University of Miami, Virginia Key, Miami, Florida. A total of 41 scientists participated.

The Workshop began with a plenary session in which the broad outlines of the multidisciplinary problems were presented. The session then dissolved into individual panel meetings. A strong effort was made at the initial session, and throughout the Workshop, to encourage discussion between chemists and meteorologists. To this end, plenary sessions were held several times during the Workshop so that progress reports could be presented by the panels for general comment and discussion. The ultimate objective of each panel was to produce a report that critically summarized present knowledge and made specific and realistic recommendations as to the types of measurements and programs that would be necessary in order to make an accurate assessment of atmospheric transport fluxes to the oceans. Rough drafts of the reports were completed by the panels during the meeting. Subsequent to the meeting, copies of these reports were circulated to all panel chairmen and other interested individuals for comments and criticisms; these were conveyed to the authoring panel. In this manner, the reports went through several stages of editing and modification before being incorporated into this present volume. The reports have not been altered in any substantive way by any persons other than the authoring panel members; thus, these documents reflect solely the opinions of the panels. This approach was necessary because of the difficulties that would have been encountered in attempting to homogenize the output of such a diverse group of disciplines. Thus, at the cost of a discontinuous style and some redundancy, we gain a directness and timeliness that would not otherwise be possible.

In conclusion, I would like to pay tribute to the participants of this Workshop. They met mornings, afternoons, and evenings for five days. Their diligence and self-discipline were all the more admirable because they had to resist the temptations of the beautiful subtropical coastal environment at the meeting site, an environment that was especially attractive in its contrast to the midlatitude winter that many had left behind. Perhaps the beauty of the ocean served as an incentive to work toward its preservation.

Joseph M. Prospero

# Contents

1. Introduction    1

2. Summary and Principal Conclusions and
   Recommendations    3

3. Modeling the Atmospheric Transport of Pollutants and
   Other Substances from Sources to the Oceans    25

4. Wet and Dry Removal Processes    53

5. Metals    124

6. Halogenated Hydrocarbons    146

7. Petroleum and Related Natural Hydrocarbons    169

8. Nonhydrocarbon Gases    181

9. Radionuclides    212

10. Techniques    222

Appendix:    The United Nations Directory of Existing
Pollution Monitoring Programs    241

xi

# 1 Introduction

The oceans, in spite of their great size and depth, are vulnerable to pollutant impacts. The effects of marine pollution are most evident near major sources such as coastal areas adjacent to urban centers. The effects are sometimes visible as a local deterioration of marine life; at other times, pollutant inputs are only detectable as anomalously high concentrations of organic and inorganic substances in organisms and in the water. Much of this pollutant material reaches the sea by direct dumping or discharges of waste. However, significant inputs can be inferred to occur via the atmosphere. For example, the occurrence of acid rains across the entire northeast United States would lead one, logically, to expect acid rains to occur off the eastern seacoast as a consequence of the westerly flow in these latitudes. More specifically, numerous studies of airborne pollutant transport and deposition patterns in the United States and Europe show well-defined relationships with specific major cities and industrial areas, often many hundreds of kilometers away. We would expect the same general transport and deposition patterns to obtain for sources of pollutants in coastal regions. However, transport from the continents to the oceans is not necessarily confined to coastal areas. There is increasing evidence that anthropogenic materials can be, and are being, transported in significant quantities to regions thousands of kilometers from their source.

Although major atmospheric inputs to the oceans can be inferred, there are few data upon which to base a valid assessment of the significance of this process. What materials are being transported to

1

the oceans, and in what quantities? Which are anthropogenic in origin, and which are naturally produced? Where and at what rates are these materials being deposited? What meteorological factors control the transport and deposition of continentally derived substances? How are the transport and deposition processes affected by the chemical and physical characteristics of the materials? This complex interrelated array of questions can be answered only through the combined and cooperative efforts of chemists and meteorologists. Indeed, the importance of the link between the chemical properties of materials emitted into the atmosphere and their effect on the physical properties of the atmosphere has become increasingly evident over the last several years. Now, in addition to our old concerns for aesthetic degradation and biological impacts, there is a new one—that man's activities may be capable of influencing weather and climate in undesirable ways; it is possible that certain anthropogenic emissions, both particulate and gaseous, could alter the radiative properties of the atmosphere and the distribution and types of cloud.

The task of the Workshop on the Tropospheric Transport of Pollutants to the Oceans as set forth in the charge of the Scientific Committee on Oceanic Research of the International Council of Scientific Unions was "to evaluate the problems involved in studying the transport of organic and inorganic particles and gases through the troposphere and their transport to the ocean, including the development of suitable sampling and analytical methods, and to consider means for promoting their investigation." This charge was indeed broad and, therefore, did not restrict the subject area of the Workshop to pollutants. In fact, the term "pollutant" is not definitive. An element or compound injected into the environment as a consequence of man's activities is not a pollutant *per se*; it becomes a pollutant when its distribution, concentration, and chemical or physical behavior are such as to have undesirable or deleterious consequences. Thus, attaching a label of pollutant to a specific material presumes a considerable knowledge about its impact on the environment, knowledge that, for the most part, is lacking for the marine environment. Consequently, the participants in the Workshop considered not only those materials that are commonly regarded as pollutants but also a wide range of other substances, both natural and man-made, including some that could serve as tracers for the study of transport and transfer processes. The objective of this approach was to define the fundamental character of these processes in a comprehensive manner so that a strategy for the solution of specific problems could be developed.

# 2 Summary and Principal Conclusions and Recommendations

## I. INTRODUCTION AND OVERVIEW

The solution of a problem of the type addressed by this Workshop requires, in a very broad sense, four types of activity: data acquisition, system modeling, monitoring, and assessment. *Data* are required to define the extent of the problem—what, where, how much, in what form, and at what rate? These data, usually obtained in specific, exploratory, limited-duration field experiments, can then be used to formulate *models,* which are needed to test hypotheses and to make projections of future trends. *Monitoring,* the routine gathering of data on a time and space scale dictated by the model, is needed to verify these projections. All three are needed to *assess consequences.*

The participants in the Workshop were asked to consider these questions in the context of their respective disciplines, to formulate conclusions, and to make recommendations. The principal conclusions and recommendations to come out of these deliberations are presented in this chapter.

The consensus of all panels was that, for most substances, the data base was insufficient to permit more than a qualitative or, at best, a semiquantitative evaluation of fluxes to the oceans. The lack of data is due to a number of factors, a major one being logistical—the oceans, which comprise 75 percent of the earth's surface, are difficult and expensive to explore. Also, the number of individuals and groups involved has been small and grossly inadequate for the task. Con-

3

sequently, the data base for most substances consists of a few measurements made in a limited number of locations over a limited time span. A second problem has to do with the quality of the data. The concentrations of many pollutants and trace substances are extremely low, and measurements require advanced state-of-the-art techniques. Also, the low ambient concentrations increase the possibility of contamination during sampling and during analysis. A third problem derives from the diversity of techniques used for sampling and analysis. Often, these techniques can yield answers that are not directly comparable.

The development of transport and deposition models calls for close cooperation between chemists and meteorologists. Consider, first, the atmospheric transport process as seen from the meteorological perspective. In the simplest sense, it can be regarded as occurring in two different modes: offshore and long range. In the offshore mode, the transport of pollutant air parcels takes place in the turbulent, relatively shallow boundary layer, which is well mixed in the vertical plane; this flow is essentially two dimensional in the horizontal plane, and the resulting deposition pattern extends directly from the source. In contrast, the long-range transport involves mechanisms that draw up the boundary layer and incorporate it into the "free" upper troposphere; here, the parcel is transported relatively rapidly until it eventually undergoes large-scale descent and again becomes incorporated into the boundary layer, either over the continents or over the oceans. During the ascending phase of the transport of the air parcel, interactions between the trace constituents and the forming, growing, and sedimenting hydrometeors can physically alter, effectively remove, or vertically displace some of the trace constituents. Because of the extreme complexity of these interactions with hydrometeors, it is difficult to diagnose and predict the three-dimensional trajectories of trace gases and aerosols moving from the continents to the oceans and, also, the physical and chemical transformations that might occur en route.

There exist a number of atmospheric (meteorological) models that may eventually be applicable to the transport problem. The suitability of these models will depend on the time and space scales of the transport process under consideration and the complexity of the chemical systems of interest. However, *it is clear that a comprehensive model capable of predicting the transport of reactive gases and aerosols is not close to realization.* Before the transport process can be modeled, we must have a much better knowledge of the chemical and physical characteristics of the materials involved, for example, reaction rates under ambient conditions; the concentration in the solid, vapor, and liquid states; the size distribution, morphology, and sorptive characteristics of aerosols; and the vertical, areal, and temporal variability of these parameters. This, and other, information is required before detailed microphysical, photochemical, and cloud models can be developed. Also, laboratory and field experiments will be necessary to determine the most important reactions so that the complexity of the parameterization can be reduced to a more tractable level.

*We have not attempted to assess possible impacts of tropospherically transported pollutants because we believe that there are insufficient data on which to base such an assessment.* Consider, for example, the problem of estimating the relative importance of nearshore (boundary-layer) transport as compared with long-range transport. On the basis of our knowledge of pollutant deposition around urban areas, we might expect that the highest rate of deposition for many types of pollutants during offshore transport would occur on waters relatively close to the major coastal sources. This close-in deposition could have a high impact because biological activity in the oceans is greatest over the continental shelves, which comprise less than 10 percent of the world oceans. However, in some coastal areas, the inputs of certain types of pollutants by rivers, coastal runoff, and waste discharge may far outweigh the atmospheric input. Thus, assessment of the relative importance of the atmospheric transport to coastal waters will require considerable knowledge of the nature and magnitude of these other types of inputs and of their transport and dispersion from the source site.

On the other hand, the transport of pollutants to relatively remote oceanic areas may take place most effectively via the atmosphere. The low velocities of ocean currents, the large mixing volumes, and the effects of scavenging could result in a large reduction in the concentration of materials injected directly into the oceans along the coasts. In contrast, materials transported via the atmosphere and injected at the sea surface are often concentrated at the air–sea interface in organic films. These films are eventually compressed and collapsed to form particulate matter, which could enter into the food chain. For these reasons, the panels concluded that investigations of both nearshore and long-distance transport and deposition should be energetically pursued.

One difficulty in assessing the magnitude of the tropospheric transport of pollutants to the oceans is ascertaining which components are, indeed, anthropogenic in origin and which are natural. This is a major problem for the study of many trace substances. To cite one example, the Metals Panel evaluated data gathered from a number of coastal and remote locations; they concluded that the anomalously high concentrations found for many metals (some of which, such as zinc, mercury, arsenic, selenium, and cadmium, are commonly regarded as pollutants) could be attributable in part or in whole to natural processes such as volcanism, low-temperature volatilization from rocks and soils, biogenic conversion to volatile species, and recycling from the oceans' surface. The same problem exists with organic materials. The problem is particularly severe with petroleum hydrocarbons, which are, in every sense, a natural product; the anthropogenic impacts derive primarily from the acquisition, transportation, and utilization of petroleum. Thus, the extent and magnitude of natural processes must be established before we can assess the anthropogenic impact.

## II.  SUMMARY OF PANEL REPORTS

The reports of the panels are summarized in this section. The summaries of the Modeling Panel and the Wet and Dry Removal Panel have been combined because of the strong overlap in subject matter. Also, in the interest of coherence, some of the conclusions from the panels dealing with substances have been incorporated, where germane, into the summary on modeling and removal.

### A.  MODELING THE ATMOSPHERIC TRANSPORT OF POLLUTANTS AND OTHER MATERIALS FROM SOURCES TO THE OCEANS; WET AND DRY REMOVAL PROCESSES

There are four basically different types of model of atmospheric flow currently in use that could be applied to the modeling of pollutant transport. *Diagnostic models* can be used to trace the three-dimensional trajectory of the center of mass of a pollutant (and hence its large-scale dispersion) either forward in time from a source to a receptor site or backward in time from a receptor site to the source. For the most effective application of this type model, sample integration times should be less than 1 hour. *Numerical weather-prediction models* are best suited for predicting transport for periods on the order of days for situations for which high temporal resolution is not required. The grid distance of 350 km imposes an effective 5–10 hour filter on predictions; consequently, sample integration times up to about 6 hours can be tolerated. *General circulation models* are most appropriate for long-term averages—weekly or monthly means. These models should be useful for predicting global tropospheric distributions of materials that have continuous sources on the continents and for predicting the mean fluxes to the ocean surface. Sample integration times should be a few days to a week. *Two-dimensional mean models* are suited to long-range predictions of seasonal means on a global scale. Sample integration times can be as long as months or seasons.

These models are currently applicable to the study of the transport of relatively inert gaseous materials but *not* to the modeling of the transport of aerosols and reactive gases. The inclusion of aerosols and reactive gases increases the difficulty of the problem enormously. In such cases, the model must account for a complex array of processes, among which are gas-to-gas and gas-to-particle conversion; particle agglomeration; wet removal through nucleation, by dissolution of soluable gases and by particle capture; dry removal of aerosols and gases to surfaces—both land and sea. Here, clouds play an especially important and sometimes paradoxical role. Cloud droplets can serve as efficient scavengers of gases and aerosols; if precipitation develops from the cloud, the captured materials are removed from the atmosphere and deposited to the earth's surface. However, only about 10 percent of all clouds actually precipitate—the remainder simply age and dissipate, leaving behind a residue of aerosol particles generated by the evaporation of the cloud droplets. Thus, nonprecipitating clouds can act as "reaction vessels" for reactive gases and as "pumps" that transfer material from the boundary layer through the inversion into the middle and upper troposphere. Although cloud microphysics and cloud dynamics are fields of active research, and although great progress has been made in understanding these processes, much more work will be required; this is especially true with regard to the behavior of the anthropogenic materials, as there is evidence that these materials might have a profound affect on cloud growth and on cloud type and distribution. Thus, we cannot model the transport and removal of pollutants without being able to model clouds, and we cannot model clouds until we know how pollutants interact with water vapor and cloud droplets.

The calculation of fluxes to the oceans will require much more information on rainfall parameters, especially composition and rainfall amounts. Most of our present knowledge about precipitation scavenging is derived from the study of radionuclide deposition. These studies indicate that the deposition due to wet processes is about five times that due to dry removal. However, it is questionable if the radionuclide results can be applied to pollutants in general—these two classes of materials have different source functions, and the particle size distributions are different, as are the chemical properties of the materials. These differences could affect not only the rates of deposition to the surface but also the relative efficiencies of the wet and dry removal processes. Moreover, most precipitation data have been obtained on the continents. There are some data from islands and from ships, but there can be a significant site-specific biasing of samples collected at such locations. Even such critical parameters as the quantity and distribution of precipitation over the oceans are poorly documented. Consequently, estimates of global deposition based on extrapolations of measurements made on land are highly questionable.

Assuming that areally representative oceanic deposition sampling sites can be found, the collection of representative precipitation (and, hence, the calculation of accurate wet deposition rates) is not difficult, providing that an automatic collector is used (i.e., a collector that opens only when precipitation occurs). However, no equivalent device exists that is suitable for the collection of the dry deposition component (that is, all material that is deposited by processes other than precipitation). This is obviously true for gases, but it also applies to aerosols. Dry deposition rates are a function of a large number of site-specific parameters, the most critical being the nature of the deposition surface. For aerosols, the collection efficiency of many of the conventional "dry" collectors is highly variable, depending on aerosol type and size; any agreement between the measured dry deposition rate of particulate matter and the true rate is purely coincidental.

Indeed, particle dry deposition is probably the area in which our knowledge of pollutant transfer to the ocean is most limited. Indicative of this is the fact that estimates of the deposition velocity for $0.1$-$\mu$m particles range over three orders of magnitude, from $10^{-2}$ to $10$ cm sec$^{-1}$. In contrast, our understanding of the physical processes governing the dry deposition (or exchange) of gases across the air–sea interface is somewhat better. In modeling the dry removal processes, it is convenient to think of the atmosphere and the oceans as being separated into layers whose boundaries are defined on the basis of the dominant transport mechanism that obtains over any specific depth/altitude region. For these layers, a characteristic transfer velocity (or its inverse, transfer resistance) can be assigned. In this manner, the rate-limiting layer(s) can be identified for specific types of materials. For example, under most conditions, for low-molecular-weight gases that are perfectly absorbed, the resistances of all atmospheric layers are comparable in magnitude and yield deposition of velocities of about $1$ cm sec$^{-1}$; however, for high-molecular-weight gases, it appears that transfer is limited by Brownian diffusion across the viscous sublayer in the atmosphere. (The viscous sublayer would also be rate-limiting for perfectly "absorbed" particles in the size range 0.001 to 0.1 $\mu$m; for larger particles, inertial effects become increasingly important and the resistance decreases with increasing wind speed.) Thus, for highly soluble gases or for gases that react in the oceans to form relatively nonvolatile products, transfer is gas-phase controlled; in contrast, for unreactive gases or gases having low solubility in water, the transfer is water-phase controlled. Using these concepts and measured air–sea concentrations, fluxes have been calculated for a number of important gases, natural and anthropogenic: $SO_2$, $N_2O$, $CO$, $CH_4$, $CCl_3F$, $CH_3I$, and $(CH_3)_2S$. The values obtained, although crude, appear to be reasonable.

In order to refine these calculations, we need more accurate values for the gas- and liquid-phase transfer velocities and more and better quality measurements of gases in both the atmosphere and the oceans; indeed, the dearth of good measurements is probably the most important factor limiting the accuracy and representativeness of these calculations. Micrometeorological techniques for the measurement of water-vapor fluxes (e.g., eddy correlation and the profile method) might be useful for the direct determination of the fluxes of gases whose air–water exchange is controlled by atmospheric (gas-phase) processes. When it comes to estimating values for the liquid-phase gas-transfer processes, the radon deficiency method appears to be most useful; this method is based on the decrease of the $^{222}Rn/^{226}Ra$ ratio in near-surface waters due to $^{222}Rn$ loss to the atmosphere. In both cases, the effects of wind and waves on transfer velocities are poorly understood and warrant considerable study.

Thus, in dealing with the problem of transport from sources to the oceans, we must be concerned with a whole subset of problems both physical and chemical in nature. Because of the numerous uncertainties and unknowns and because of the many parameterizations and simplifications that inevitably will be incorporated in any pollution transport model, *these models will require validation.* There are a number of materials, both natural and anthropogenic, that could be useful as tracers for this purpose. These include the chlorofluorocarbons, carbon monoxide, carbon dioxide, and $^{222}Rn$. These materials are gaseous and relatively inert, and they are emitted exclusively, or primarily, from the continents; thus, the modeling problem will be considerably simplified. There are a number of programs that are concerned with the study of these (and other) materials in the environment. The data output from such programs could be useful for validating transport models. *However, in many programs, including most of the routine monitoring programs, the frequency of sampling and the duration of individual sampling periods are such as to preclude the use of the resulting data for model validation purposes.*

B. METALS

There is general agreement that the major portion of the aerosol mass over the oceans is derived from two sources: the sea itself, as spray from bursting bubbles, and the earth's crust, as mobilized soil material. However, the concentration of a number of elements in this material is much greater than would be expected on the basis of the average composition of crustal material (soils or rocks) or of seawater. This suggests the possibility that these anomalously high concentrations could be due to anthropogenic inputs. However, the enrichment factors for these elements in aerosols (relative to crust or seawater) are relatively independent of geographical location, being approximately the same, for example, over the North Atlantic and at the South Pole even though the aerosol concentration is a thousand times lower at the latter site. This similarity in enrichments would seem to indicate (but does not prove) that natural processes might be responsible. However, there are few data on the rates of mobilization of materials to the atmosphere by natural processes other than the generation of soil aerosols; even the latter is not well known. A crude calculation of metal fluxes to the atmosphere can be made on the basis of published estimates of soil dust and sea-salt aerosol fluxes and the composition of emissions from fossil-fuel combustion. Such a calculation suggests that the flux of Pb, Hg, and Se from fuel combustion dominates that from natural sources by at least a factor of 10, while the fluxes of V and As from these two sources are of comparable magnitude; these heavy metals are commonly regarded as pollutants.

Because of the dearth of data on aerosol composition over the oceans and because of the complete lack of data on the composition of precipitation, it is difficult to estimate deposition to the world's oceans. A rough estimate can be made for the North Atlantic, where there is some semblance of an aerosol data base. Using these data and a rain statistic based on the mean residence time of water vapor in the troposphere, the elemental fluxes to the oceans are obtained. These fluxes can then be compared to elemental deposition rates to the ocean floor as calculated from measured sedimentation rates and the known composition of sediments; the latter rates are, in effect, long-term averages of over a thousand years and, thus, would not be greatly affected by recent anthropogenic inputs. This comparison indicates that the atmospheric inputs of Pb, Hg, and Se are anomalously high along with Cd, Sb, and possibly Zn. The high (calculated) input rates could indicate an anthropogenic origin; however, alternative explanations are (1) that material is being recycled from the sea surface to the atmosphere and/or (2) that the removal efficiency for small particles (which have very high enrichment factors) is much less than for larger particles. There is some evidence to support all these possibilities; however, it is impossible to determine at this time which of these, or other, factors might be responsible for the anomalously high relative concentrations (or enrichments) of certain metals in aerosols or the high calculated atmospheric input values to the oceans, at least to the North Atlantic.

The effect of anthropogenic activities on the mobilization of metals (and other materials) and on the input of these materials to the oceans could be readily assessed if we had available a chronological record of atmospheric deposition that predates the era of heavy industrialization. Such a record may be obtainable from marine sediments and glaciers. With sediments, adequate time resolution could be obtained only in regions of very high deposition rates such as those found in some nearshore environments. Glaciers are ideal because of the high snow accumulation rates and the relative ease with which a chronology can be established. However, both approaches have a number of shortcomings that can be overcome only by further work and the development of better techniques.

C. HALOGENATED HYDROCARBONS

Two classes of halocarbon are of special interest: low-molecular-weight (LMW) halocarbons, primarily because of their potential impact on the ozone layer, and high-molecular-weight (HMW) halocarbons, because of their possible deleterious effects on marine organisms. The general consensus is that, at this time, there is no recognized direct impact of either group of materials on the oceans although the bioaccumulation of the HMW species has been established, and possible indirect effects via bioaccumulation in the food chain have been postulated in specific cases. However, the data available on the distribution of these materials in the oceans and the atmosphere are very limited.

There are major natural as well as anthropogenic sources for the LMW halocarbons. The oceans appear to be the primary source by far for the methyl halides ($CH_3Cl$, $CH_3Br$, and $CH_3I$). There is a second group of LMW halocarbons whose major source appears to be anthropogenic: vinyl chloride, ethylene dichloride, trichloroethylene, and perchloroethylene. These have very short lifetimes in the atmosphere (hours to one day), and, thus, transport to the oceans via the atmosphere is not expected to be important. However, these compounds have been measured in the atmosphere over the open oceans and in the oceans themselves. Moreover, these materials have been detected in drainage water, and, thus, rivers may be the main route to the oceans. Thus, the source of these halocarbons in remote marine environments has not been clearly established.

The third group of LMW halocarbons consists primarily of the one- and two-carbon chlorocarbons and fluorocarbons. The chlorofluorocarbons (or Freons), because of their low reactivity in the atmosphere, should have long residence times, and, indeed, they are found to be relatively uniformly distributed throughout the troposphere within the northern and southern hemispheres (although the mean concentration in the northern hemisphere is appreciably greater). The LMW chlorocarbons $CCl_4$, $CHCl_3$, $CH_2Cl_2$, and $CH_3CCl_3$ have been measured in significant concentrations in air over the North Atlantic Ocean. However, many of the LMW chlorofluorocarbons and chlorocarbons have also been detected in river systems, and, thus, river transport may be significant for some compounds.

The HMW chlorinated hydrocarbons of primary interest are the pesticides such as DDT and the class of compounds known as the PCB's. It seems clear that, because of the low water solubility of these materials, the major mode of transport must be through the atmosphere. Fairly extensive measurements of pesticides have been made in continental air, and some attempts made to measure them over the oceans; the latter efforts are hampered by the difficulty of making measurements at the very low levels that appear to obtain. However, data suggest that the marine air concentrations are at least one to two orders of magnitude lower than over the continents.

As for ocean waters, there is much more information on PCB and pesticides in marine organisms than in the water itself. Indeed, no one has reliably measured DDT in the open ocean except in surface films of organic matter, where DDT (and also PCB's) are enriched relative to the underlying water. However, PCB's can be measured directly in ocean water, and a very marked (factor of 10) reduction was noted in the North Atlantic between 1971 and 1973; the sharpness of the reduction has not been satisfactorily explained but has been attributed to the combined effects of the world sales restriction and scavenging by sinking particles.

There are few data for pesticides and PCB's in marine environments in the southern hemisphere. However, it is safe to conclude that the predictions, made in the early 1970's, of large increases in atmospheric and water concentrations throughout the world have not been realized.

We can only speculate on the primary modes for the removal of the halogenated hydrocarbons from the atmosphere. For the LMW halocarbons, the principal path is through chemical transformation, possibly with the OH radical playing an important role. Gaseous exchange at the air–sea interface may be significant, but wet removal and aerosol deposition are not. For the HMW chlorocarbons, model calculations suggest that gaseous exchange, wet removal, and aerosol deposition could all be significant; however, there are virtually no data to support this assertion.

## D.  PETROLEUM AND RELATED NATURAL COMPOUNDS

Crude oil is a complex mixture of thousands of individual compounds, mostly hydrocarbons representing several homologous series. Once this material is injected into the atmosphere, it is subject to chemical reactions that lead to simple oxidation or to polymerization reactions that ultimately result in particle formation. Because of the varying degree of reactivity of the constituent organics and because of the variable composition of petroleum and also because of the lack of sufficient kinetic data, it is difficult to estimate the rates of conversion in the atmosphere. These rates are important because they affect the lifetime in the atmosphere of individual species and, hence, the probability of transport of that species from the continent to the oceans. Also, the rate of conversion could affect the rate of removal of organic materials from the atmosphere. The oxygenated (more polar) reaction products would, in general, be more soluble, a factor that would increase the efficiency of wet removal or of gaseous deposition at the sea surface; conversion to particulates would also increase the efficiency of wet and dry deposition.

However, the existing data on the sources, concentration, and composition of hydrocarbons in the atmosphere is sparse, confusing, and often contradictory. Estimates of the anthropogenic emissions to the atmosphere (excluding methane) range from $45 \times 10^{12}$ to $68 \times 10^{12}$ g $yr^{-1}$, of which $30 \times 10^{12}$ to $65 \times 10^{12}$ g $yr^{-1}$ is believed to be transported in the atmosphere to the oceans. The estimates of the transport of natural hydrocarbons (nonmethane) to the ocean are even more uncertain and range from 0 to $220 \times 10^{12}$ g $yr^{-1}$.

In addition to the continental source of natural hydrocarbons, the oceans act as a source. However, the organic material entering the atmosphere from oceanic sources will be composed primarily of volatile, gas-phase compounds that are highly paraffinic and stable and, thus, are not likely to be rapidly converted into particles that would then be redeposited to the sea. On the other hand, a significant fraction of the anthropogenic hydrocarbon materials injected into the atmosphere over the continents is reactive and is converted to products that enter the aerosol phase; it is estimated that $0.6 \times 10^{12}$ to $1.4 \times 10^{12}$ g yr$^{-1}$ of this material is deposited in the ocean. In contrast, the flux of light organics from the oceans to the atmosphere is estimated to be about $1.4 \times 10^{12}$ g yr$^{-1}$, and the total atmospheric burden of organics over the oceans, about $11 \times 10^{12}$ g.

Wet removal is expected to be important for many classes of compounds. However, there are virtually no data on the types and concentration of hydrocarbons in precipitation in marine areas (or even in nonurban continental regions).

### E.  NONHYDROCARBON GASES

The materials considered in this category are various gaseous species of carbon, nitrogen, and sulfur. It is clearly recognized that these gases cannot directly alter the chemical properties of the ocean in any significant way. These gases were considered because they can react in the atmosphere with water, ozone, and their related free-radical species to form products that could have a profound effect on climate. The central species is sulfur dioxide, which is emitted as a pollutant, mainly over the continents, and which becomes oxidized to sulfuric acid and sulfate in the atmosphere. These end products form aerosols that could have a direct influence on climate by altering the radiation balance of the earth. Also, because of the hygroscopic nature of sulfates, these aerosols could influence climate indirectly by affecting cloud microphysical process and dynamics, thereby changing the albedo. The other species considered were those that might significantly affect the rate of conversion of $SO_2$ to particulate sulfates. Man's contribution of $SO_2$ to the global flux of atmospheric sulfur compounds is estimated to be 35 to 45 percent. The present rate of growth in the use of fossil fuels containing sulfur is such that anthropogenic $SO_2$ emissions will soon dominate the global cycle of sulfur through the atmosphere. In addition, the increased use of tall stacks to aid in the dispersion of $SO_2$ emissions (so that local pollution control regulations are not violated) will increase the quantity of $SO_2$ and sulfate particulate matter injected above the boundary layer; this will, in effect, increase the efficiency with which these materials are transported to the ocean environment. Another factor of possible importance is that the increasing $SO_2$ emissions may consume $NH_3$ at such a rate as to cause a significant reduction in $NH_3$ concentrations in the atmosphere. This would serve to increase the residence time of $SO_2$ in the atmosphere and, thereby, increase transport to the ocean.

In the global cycles of sulfur, nitrogen, and carbon compounds, the oceans can act either as a source or a sink (or both). However, these cycles are, for the most part, poorly understood. For example, the strength of the global source of biogenic sulfur has been estimated to range from $32 \times 10^{12}$ to $110 \times 10^{12}$ g of S $yr^{-1}$; estimates of the oceanic contribution to this flux range from $27 \times 10^{12}$ to $50 \times 10^{12}$ g of S $yr^{-1}$. These large discrepancies are due to the fact that the source estimates are derived indirectly; they are simply the fluxes required to make the world budget balance. However, many of the source strengths used in the budget are based on a few isolated data and require further substantiation.

As stated previously, we must be concerned with those species and reactions that affect the rate of conversion of gaseous sulfur to the particulate phase. One of the most important reactions appears to be that of $SO_2$ with the hydroxyl (OH) radical to produce $H_2SO_4$. The OH radical is highly reactive with a number of atmospheric gases and has an average lifetime of about 1 sec. Because of its reactivity and because it is produced through reactions of ozone, water vapor, and ultraviolet radiation, we can expect the OH concentration to depend strongly on altitude, latitude, and solar angle. Despite its importance, only one set of measurements of ambient OH concentrations has been reported to date.

Aside from OH, the gases that have the greatest effect on the conversion of $SO_2$ are those that affect the steady-rate concentration of OH. Important gases of this type are ozone, hydrocarbons (natural and anthropogenic, but especially methane), oxides of nitrogen, and carbon monoxide. Thus, long-range predictions of sulfate aerosol formation could be very much dependent on long-term changes in the concentrations of many other pollutant gases. Carbon monoxide is especially important in this regard for, already, northern hemisphere concentrations are three times those in the southern hemisphere.

The manner in which these gases react in the atmosphere to influence $SO_2$ conversion is poorly understood. Reaction mechanisms are often subject to debate, and the role of nucleation processes and aerosols in the reaction steps is not clear. Rate constants are often of questionable quality. Also, the essential concentration data are fragmentary and often totally lacking in certain critical areas such as the southern hemisphere. Such information must be obtained before transport to the oceans can be modeled in any acceptable manner.

## F. RADIONUCLIDES

The principal sources of the man-made radionuclides in the atmosphere are the detonation of nuclear devices and releases from operations in the nuclear-power fuel cycle (mining, fuel fabrication, reactor operation, fuel reprocessing, and waste management). The panel has excluded from consideration the fate of radionuclides directly emitted in weapons tests and the analysis of the consequences of catastrophic incidents such as a nuclear-reactor excursion; these subjects have been thoroughly studied and reported on in the literature.

The principal radionuclides emitted in routine reactor operations are $^3$H (tritium), $^{131}$I, and $^{85}$Kr. At present, approximately 95 percent of the atmospheric inventory of tritium is weapons-produced, and the balance is essentially all derived from reactors. There is no evidence that present or projected ambient tritium concentration levels could have any significant effect on ocean physical or biological systems. Likewise, $^{131}$I, because of its short half-life, should not be transported to the ocean in significant quantities. As for $^{85}$Kr, 95 percent of the present atmospheric burden is derived from the nuclear power cycle and 5 percent from weapons. There is some concern that the projected increase in the atmospheric concentration of $^{85}$Kr (through the increased reliance on nuclear power) could lead to a significant (43 percent) increase in air ionization by the early part of the twenty-first century. There has been some speculation that the increased ionization would have a perceptible impact on a number of meteorological processes. However, at this time, there is no firm evidence linking $^{85}$Kr to atmospheric phenomena.

There is some question as to the source reservoir of the weapons-produced radioactive material in the lower troposphere. How much is "young" material derived directly from the stratospheric reservoir, and how much is "old" material that had been previously deposited on soils and subsequently resuspended? A high ratio of resuspended ("old") to stratospheric ("young") material would imply that air concentrations are decreasing, and would continue to decrease, at a lower rate than the decrease in the rate of injection through testing, i.e., the soils would act as a long-term reservoir for atmospheric material and, thus, at some point, "fallup" could dominate "fallout." However, calculations performed for a worst-case situation (plutonium in soils and air at Rocky Flats, Colorado) suggest that resuspension can only account for several percent, at most, of the airborne radioactive material. Therefore, it appears that the resuspended input to the atmosphere will be essentially undetectable until all atmospheric testing has stopped and the stratospheric reservoir has been largely depleted.

The natural radionuclides found in the atmosphere are primarily members of the uranium decay series, in particular, $^{222}$Rn and its radioactive decay (daughter) products. The flux of radon from soils is approximately 100 times greater than that from the oceans. Thus, radon, because of its unique source characteristics and its chemical inertness, would be an ideal tracer for the validation of atmospheric transport models for nonreacting species. However, there are many aspects of the source strengths, atmospheric concentrations, and depositions of radon and/or its daughters that are not well characterized and that will require further study before these nuclides can be used for this purpose.

G. TECHNIQUES

Over the past decade, there has been a great increase in the number of types of aerosol collection instruments in use. These include instruments that perform a size classification function, such as impactors. In many cases, the sampling efficiency of these instruments under ambient conditions (or their laboratory equivalent) has not been determined. For example, only recently has it been learned that the standard high-volume air-sampling system used in the United States has intake characteristics that result in a sharply reduced sampling efficiency for particles above 5 $\mu$m; also, the intake efficiency is strongly dependent on wind velocity and the sampler orientation with respect to wind direction. As for impactors, recent developments in theory have made it possible to design instruments with optimal operational characteristics for ideal (test) aerosols; however, with real aerosols, problems arise because of particle shattering and bounce off, re-entrainment, stage-loading effects, and humidity effects. Because of the uncertain performance characteristics of various instruments and devices, it is often difficult to make meaningful comparisons between data sets obtained with different instruments.

The major problems in attempting to characterize the chemical composition of precipitation are related to the collection of quantitative samples and to chemical reactions during storage. At *land* stations, representative precipitation samples can be obtained with devices such as the automatic wet–dry collectors, which are only open during the time that precipitation actually is falling. Samples collected by devices that are open at all times (i.e., bulk, or total precipitation, collectors) are of questionable usefulness; the aqueous phase cannot be assumed to be representative of precipitation because of reactions between the water and the particulate "dry-deposited" component, and the "dry" component cannot be assumed to be representative of actual dry deposition. Indeed, as stated previously, there is no satisfactory way in which to sample the dry deposition component. In contrast to the situation on land, *the collecting of precipitation samples of suitable quality and quantity at sea remains a major problem for which no immediate solution is apparent.*

To assess the flux of materials to the oceans, it will be necessary to have information on the vertical distribution of concentrations as well as on the areal distribution. At present, these data must be obtained by aircraft; because of the high cost of aircraft time, and because of operational limitations, we cannot expect to develop an extensive data base by this means. However, at present, there is available a relatively simple, hand-held, inexpensive device—the sun photometer—that can provide information on the vertically integrated concentration of aerosols in the atmosphere, i.e., the atmospheric turbidity. Routine measurements are presently made in a network of stations in the United States and in the World Meteorological Organization monitoring program. With these data, large-scale patterns of turbidity could be related in some cases to major natural and anthropogenic sources. However, the accuracy and stability of these instruments and the quality and quantity of the observations must be greatly improved before the data from these devices can be consistently relied upon. Also, the sensitivity must be improved for use in remote locations where the turbidity is low and changes are small.

## III.  CONCLUSIONS AND RECOMMENDATIONS

The following statements summarize the conclusions and recommendations of the Workshop panels:

A. THE PRINCIPAL EFFORT in the investigation of the tropospheric transport of pollutants to the ocean must continue to be directed along research lines. At present, there are insufficient data with which to design a sound monitoring program for most substances. In addition, the shortcomings of the sampling and analytical techniques employed for many of the trace substances under consideration make them unsuitable for routine monitoring in most oceanic regions where ambient concentrations are very low. The use of these procedures under such conditions in a monitoring mode could lead to severe quality-control problems. It would be more desirable to have a limited set of high-quality data than a large set of doubtful data.

B. THERE IS A GREAT NEED for more concentration and composition data in marine regions both in the atmosphere and in the ocean. Specific needs are listed by substance.

*1. Metals*

*In the Atmosphere* To assess sources and fluxes, we need measurement of the elemental and chemical composition of aerosols, of particle size distribution and, especially, of particle composition as a function of size.

*In the Oceans* To assess deposition and recycling rates, the concentration in water and in surface slicks is required for both the dissolved and particulate phases.

*Over the Continents* Investigations should be undertaken to determine the major sources of metals in the atmosphere. Emissions from anthropogenic sources such as smelters, cement plants, oil, and gas power plants must be measured and characterized on a global basis. Important parameters are elemental (or chemical) composition, particle size distribution, and particle composition as a function of size. Emissions from specific natural sources and processes such as volcanism, the biosphere, low-temperature volatilization and crustal degassing, and chemical fractionation at the air–sea interface must be evaluated in the same manner.

*2. Halogenated Hydrocarbons*

*In the Atmosphere* Especially lacking are measurements of the high-molecular-weight chlorinated hydrocarbons in relatively remote ocean areas. The concentrations in such regions are very low, often below present-day detection limits. Thus, such measurements will require the development of better instrumentation and improved techniques to minimize the possibilities of losses and contamination. It is extremely important that measurements be made simultaneously of both the vapor-phase and the particulate-phase concentrations and of the concentration in precipitation.

*In the Oceans* In order to assess the global fluxes to the ocean from the atmosphere and from rivers, the concentrations fields in the surface (mixed) layer and vertical profiles across the thermocline must be well characterized. The concentration in phytoplankton and zooplankton should also be measured to determine the degree of biomagnification and the extent of natural biological production. For those compounds that have both natural and anthropogenic sources, the carbon-14 activity might be useful for determining the anthropogenic component and, therefore, should be measured. Intensive studies should be made in waters along the Gulf Coast, where many major manufacturers of halocarbons are situated.

### 3. Petroleum and Related Natural Hydrocarbons

*In the Atmosphere* Emphasis should be placed on the hydrocarbons of high molecular weight rather than on the gaseous low-molecular-weight compounds, as it is likely that the latter will not contribute significantly to the net flux from the air to the sea. Scientists should be encouraged to obtain detailed spectra of the individual hydrocarbons by using state-of-the-art instrumental techniques; these data will make easier the task of determining the origin of the hydrocarbons, especially with regard to the distinction between biogenic and petroleum sources. A greater effort must be made to characterize the areal and vertical distribution fields of gas-phase and particulate-phase hydrocarbons.

The reaction kinetics of hydrocarbons in the atmosphere must be better understood if we are to determine the lifetime and fate of the various hydrocarbon classes. Important areas of study are gas-to-particle conversion and the distribution coefficients between solid and gas phases for various hydrocarbon species.

### 4. Nonhydrocarbon Gases

*In the Atmosphere* To attain a better understanding of the global sulfur cycle, and to assess the possible climatic impact of sulfate aerosols, more measurements are needed of $SO_2$, $H_2S$, $CH_3SH$, $(CH_3)_2S$, and those gases that are important to the conversion of sulfur species to particulate sulfate: NO, $NO_2$, $HNO_3$, $NH_3$, $O_3$, OH, $H_2O_2$, and CO. In the aerosol phase, sulfate and nitrate must be measured.

Because of the high degree of interrelatedness among these species, it is essential that experimental programs be designed to maximize the number of species measured simultaneously.

Also required will be theoretical and laboratory studies of the following processes and reactions: (1) the rate constants for the elementary homogeneous chemical reactions of the above-mentioned gases, (2) the formation of nuclei by heteromolecular chemical reactions, (3) the heterogeneous oxidation of $SO_2$ in solutions, (4) the heterogeneous oxidation of $SO_2$ in gas–solid systems relevant to the atmosphere.

## 5. Radionuclides

*In the Atmosphere* The distribution of radon-222 and its daughter products, especially lead-210, must be better characterized over the continents and the ocean, especially the latter. The degree of attainment of equilibrium between radon and its daughter should be studied over a wide range of meteorological and climatic conditions.

*Over the Continents* The mechanisms for the mobilization of soil radioactivity (i.e., the resuspension of fallout nuclides) and the subsequent transport through the atmosphere should be more thoroughly investigated. This problem should be considered in the context of the general problem of the mobilization of soil aerosols.

## 6. General

*There is an urgent need for data in the southern hemisphere.* Because the major centers of production and consumption are in the northern hemisphere and because interhemispheric exchange rates are slow, the environment in the southern hemisphere has been relatively unaffected by pollutant inputs. However, the rate of population growth and of economic development suggests that this situation will inevitably change. Therefore, we recommend that a strong effort be made to study the present state of the atmosphere and ocean in that region because such studies will (1) provide an insight into chemical processes as they occur in a clean environment for comparison with processes in the northern hemisphere and (2) permit us to observe, over the coming years, changes in the chemical environment as a function of increased industrialization and consumption of fuels and resources.

By making such comparisons and observing progressive changes, we will be able to distinguish more clearly between natural processes and anthropogenic effects.

C. IMMEDIATE EFFORTS must be made to initiate procedures or programs that will lead to an increase in the accuracy and precision of gas and aerosol concentration and composition data. At present, there are few standard reference materials that meet the needs of environmental scientists working in the marine environment. It is essential that a mechanism be established whereby suitable standards can be formulated, certified, and distributed expeditiously. Often, developments in environmental studies have occurred rapidly. In the past few years, there have been a number of instances where concern over the dissemination into the environment of certain elements or compounds resulted in a sharp increase in field sampling programs. In many cases, the quality of these data might have been greatly improved if standards could have been made available on short notice. At present, several years are required for the preparation of reference materials, a procedure far too slow for our purposes. Consequently, we urge SCOR to explore the possibility of designating an appropriate international laboratory that would be responsible for the rapid preparation, and dissemination, of environmental reference standards and for the organization of interlaboratory intercomparison exercises.

The performance characteristics of many gas and aerosol sampling devices have been poorly characterized. We recommend that a workshop be held on the subject of collection techniques and that a field intercomparison be carried out in conjunction with the workshop or subsequent to it. Following the workshop and the field intercomparison, a set of recommendations should be compiled that specify preferred minimum design parameters for collectors.

The routine use of sun photometers in aerosol programs is to be recommended. However, these instruments in their present form suffer from a number of shortcomings, including calibration drift and a lack of sensitivity for measurements in regions where aerosol concentrations are low. A strong effort should be made to improve instrument design and to establish calibration facilities.

D. DATA ON THE REMOVAL OF RADIONUCLIDES from the atmosphere could be used to draw inferences about the behavior of aerosol particles in the transport and deposition cycle. However, before this can be done, we need more information on the character of the aerosols to which the radionuclides are attached, especially their size.

E. THERE IS A GENERAL LACK of storm precipitation statistics. The information required includes storm height, areal extent, frequency of occurrence, duration, precipitation amounts, and cloud water removal efficiency.

F. MUCH COULD be learned about scavenging efficiencies and the modification of substances within clouds by performing budget studies for progressively more complex nonprecipitating and precipitating clouds and storms. Experiments should start with simple wave and cap clouds, progress to orographic and cumulus clouds, and culminate in a study of cumulonimbus and frontal storms.

G. THE COLLECTION OF PRECIPITATION at sea remains a major problem. There are two types of difficulty: to obtain quantitative samples and to obtain samples free from contamination. A workshop should be convened on the subject of precipitation collection techniques and strategies for the marine environment.

H. LITTLE IS KNOWN about the dynamics of the air–sea interface. These uncertainties are detrimental to our obtaining an understanding of dry deposition and resuspension of particles and gases. Needed are concerted scientific studies on the turbulence above and below the interface, wave mixing, the physics of wave breaking, gas-particle entrainment, particle resuspension, and thermally driven circulations above the thermocline.

I. FOR MANY GASES, the dominant resistance to transfer occurs in the liquid phase and probably in the viscous sublayer. Wind-tunnel studies could help to establish the importance of evaporation and condensation, waves, spray, bubbles, wind speed, and surface films.

J. PARTICLE DRY DEPOSITION and resuspension are probably the most glaringly deficient aspects of our knowledge of pollutant transfer to the ocean. To remove the uncertainties, efforts on all fronts are recommended: controlled wind-tunnel studies, theoretical studies, and semicontrolled deposition experiments on ponds and lakes.

K. AN EXPERIMENT should be performed in some coastal urban region to measure the direct transport within the boundary layer to the ocean. Here the problem is reduced to one of quasi-two-dimensional transport. A coastal transport study has a number of advantages over one carried out over the continent—the topography is uniformly flat and there are no downwind sources to complicate the input term computations. On the basis of meteorological considerations and the distribution of major sources, the most logical site for such an experiment would be in the region off the northeast coast of the United States.

L. PROGRAMS SHOULD BE DEVELOPED to utilize certain materials that are currently being emitted into the atmosphere as tracers for large-scale transport model validation. Natural and anthropogenic processes generate a number of substances that could be useful as tracers. These include chlorofluorocarbons, carbon monoxide and carbon dioxide, and $^{222}$Rn. There are a number of programs concerned with the study of these and other materials in the environment. Consideration should be given to the feasibility of modifying these programs so that the data could be used for model testing. One approach may be to establish specific time periods, worldwide, when programs could switch to a suitable data-gathering mode.

M. MANY LARGE-SCALE, international meteorological programs are currently being planned, and some, such as the First GARP Global Experiment, are close to execution. In many cases, these experiments will take place in regions where there is a dearth of data on the concentrations of important chemical species in the atmosphere and in the oceans. Chemists should be encouraged to participate in these meteorological field experiments.

N. WITHIN FIVE TO TEN YEARS, it should be possible to conduct a major experiment to test modeling concepts of long-range transport from the continents to the ocean. This type of transport, being three dimensional and involving, among other processes, cloud formation and subsequent hydrometeor removal of aerosols and soluble gases, will be most difficult to study. A multiaircraft experiment will be essential, and a comprehensive array of surface monitors (ships and instrumented buoys) will also be required. Although to plan such an experiment now may be premature, it is not too early to discuss its possibility and to engender interest.

O. THE TIME TRENDS of the atmospheric transport of pollutants and other materials to remote areas should be established through studies of concentration changes with depth in glacier snow and ice. With these measurements, and with a knowledge of the historical trends of anthropogenic source terms, an assessment can be made of the anthropogenic impact on a global scale. Important components would be soil dust, volcanic debris, sea-salt components, metals, and organic materials. However, if glacier snow and ice is to be used for this purpose in a systematic fashion, it will be necessary to develop new techniques for collecting larger cores using more stringent contamination-control procedures.

# 3 Modeling the Atmospheric Transport of Pollutants and Other Substances from Sources to the Oceans

## I. INTRODUCTION

Most continentally derived material that is present in the atmosphere has been emitted from sources at the earth's surface. In most cases, this material, be it natural or anthropogenic in origin, is injected into the boundary layer, that portion of the troposphere that is in convective contact with the earth's surface. For materials emitted from sources situated relatively close to the oceans (i.e., within tens to hundreds of kilometers), the primary path of transport to the oceans may lie inside the boundary layer. Over longer distances, the major path will most likely be through the troposphere above the boundary layer.

A transport model must consider emissions from an array of individual sources imbedded in an atmosphere whose characteristics are such that the meteorological parameters may differ markedly at each source site. The model should also account for the dispersion through a boundary layer whose depth is highly variable both areally and temporally. For long-distance transport, mixing between the boundary

Members of the Working Group on Modeling the Atmospheric Transport of Pollutants and Other Materials from Sources to the Oceans were E. F. Danielsen and J. W. Deardorff, *co-chairmen*; D. Deaven, G. R. Hilst, J. D. Mahlman, E. R. Reiter, E. Robinson, and J. W. Winchester.

layer and the troposphere and the subsequent large-scale movement in the troposphere need to be predicted. Finally, transport downward through the oceanic boundary layer must be accounted for. The scales of motion encompassed by the model (either directly or indirectly) range from the molecular to the planetary waves in the atmosphere with wavelengths of thousands of kilometers. The modeling of such a transport system is extremely difficult even if it is restricted to the simplest case—the transport of a chemically inert species. However, the inclusion of chemically reactive species and the effects of wet and dry removal processes increase the complexity of the problem enormously.

When little is known about atmospheric (and oceanic) transport, it is often instructive to develop simple reservoir models. In such models, the pollutants are assumed to be well mixed within each reservoir and the exchange rates between reservoirs are assumed to be proportional to the mean concentrations. These models lead to the concept of characteristic residence times. Although such residence times are helpful in establishing reasonable orders of magnitude, they should not be taken too seriously, because the basic assumptions are only approximately valid. In the atmosphere, a broad spectrum of wave motions tends to generate laminar inhomogeneities in which the concentrations of a trace constituent can exceed the surrounding concentrations by as much as a factor of 10. These same wave motions affect the three-dimensional transport of trace constituents and interfere with the simple concept of residence times.

At present, better methods and numerical models are available, or are being developed, for computing the atmospheric concentrations, transports, and depositions of trace constituents. Transports by both mean and turbulent motions are included in these models. The former advects or moves the tracer's center of mass in two- or three-dimensional space and deforms the tracer's distribution, while the latter spreads the tracer over a larger and larger volume, thereby reducing its concentration.

In this chapter, current methods and models will be reviewed, discussed, and evaluated. Each has its own advantages and limitations. The emphasis will be on transport from continental sources to oceans, with a logical shifting from deterministic to statistical transport as the time scale increases. Removal and deposition processes are discussed in Chapter 4; only brief mention of them will be made here. However, it must be stressed that removal processes must not be regarded as merely an end result of transport; rather, removal processes are an integral part of the problem, constantly modifying the complex of species-related parameters involved—concentration, size distribution, and chemical characteristics.

## II. SOURCE IDENTIFICATION FOR TRANSPORT MODELING

There are at least three source characteristics that are important in the modeling of pollution transport:

1. The emission rate (mass per unit time) of primary pollutant material from point sources (e.g., power plants of cities) or area sources (such as cities);
2. The atmospheric concentrations of the primary pollutant(s) at all points within the model at the initial time of application of the model;
3. The atmospheric concentrations of other, secondary pollutant, materials that can react with, or be produced from, the primary pollutant(s).

In general, reactive species necessitate the inclusion of prediction equations for all secondary species of pollutants.

### A.  PRIMARY POLLUTANT SOURCES AND EMISSIONS

There have been very large expenditures of money and manpower by government agencies and private industries to estimate emissions from known sources of air pollutants. As a starting point for a modeling study of a source area in the United States, basic source emission data can be assimilated from federal, state, and local air-pollution agencies and from major industries and industrial trade associations. In areas where emissions may not have been estimated and tabulated, emissions can be calculated from data on fuel consumption, industrial production, and transportation usage, by applying available emission factors to these data. Estimates of global emissions are typically made in this manner. United Nations publications on fuels, agriculture, and industrial production are good sources of basic data for such calculations. Of course, for regional studies or short-term transport calculations, such data may be erroneous by a considerable factor. However, even greater uncertainties exist, one of which involves the atmospheric process by which new or secondary pollutant compounds may form from the precursor, or primary pollutant emissions. This process is considered in the next section.

## B. SECONDARY POLLUTANT SOURCES AND SINKS

When a model calculation incorporates pollutants that have any significant degree of reactivity, the formation of secondary pollutants must be treated. This is done by inclusion of extra terms within the "conservation" equation for species $i$ of mean concentration $\bar{C}_i$:

$$\frac{\partial \bar{C}_i}{\partial t} = -\bar{V} \cdot \nabla \bar{C}_i - \frac{\partial}{\partial X}(\overline{u'C_i'}) - \frac{\partial}{\partial y}(\overline{v'C_i'}) - \frac{\partial}{\partial Z}(\overline{w'C_i'}) - \bar{C}_i/\tau_{ri}$$

$$- K_{ij}(\overline{C_iC_j} + \overline{C_i'C_j'}) + K_iK_{lm}(\overline{C_lC_m} + \overline{C_l'C_m'})$$

$$+ \text{ other physical sources}$$

$$- \text{ other physical sinks.} \qquad (3.1)$$

The first term on the right represents the mean advection or transport of the pollutant; the next three terms represent turbulent transports occurring on a scale smaller than the averaging volume represented by the overbar symbol; $K_{ij}$ and $K_{lm}$ represent reaction constants between species $i$ and $j$ that decrease $\bar{C}_i$ or between species $l$ and $m$ that form species $i$. In addition, three-body reactions may also be considered as well as fluctuation terms associated, for example, with temperature dependence of the reaction rate. These source or sink terms, involving the reaction-rate constant, contain subgrid-scale components such as $\overline{C_i'C_j'}$ that indicate the degree of "unmixedness" of two constituents within the model grid volume. If $C_i$ and $C_j$ happen to be totally unmixed, yet are within the same grid volume, the value of $\overline{C_i'C_j'}$ turns out to be equal but opposite to that of $\bar{C}_i\bar{C}_j$ so that no reaction can occur. Prediction equations may be written also for these subgrid-scale concentration covariances (see Donaldson, 1973, or Hilst, 1973) in which still higher turbulence moments appear for which closure assumptions are necessary.

A distributed and natural surface sink and/or source appears within the fourth term on the right. Upon use of finite differences, as in a box model, the term $(\overline{w'C_i'})_0$ appears, where the subscript 0 refers to evaluation of the flux at the surface. The parameterization of this term is the topic of Chapter 4.

For a radioactive material, $\tau_{ri}$ in Eq. (3.1) is the half-life that is otherwise infinite.

Another process that is important in transforming the primary pollutants is the agglomeration of very fine particles into an aerosol of larger size so that the aerosol may subsequently be deposited to the surface more quickly. At present, there is no satisfactory way to model an aerosol size spectrum in a sufficiently simple manner to allow its inclusion within a three-dimensional transport model.

### C. SOME ''GENERAL'' POLLUTANT CATEGORIES; TRANSFORMATION AND REMOVAL PROCESSES

In pollution-transport modeling, pollutants are subjected to a variety of physical and chemical processes that transport, dilute, transform, and scavenge the pollutants as described in Eq. (3.1). For purposes of discussion, we may divide pollutants on a physical basis into two general classes: gases and particles (see Figure 3.1). The particles may be solids or solution droplets whose radii can extend over several orders of magnitude.

Gaseous pollutants are classed as either "nonreactive" or "reactive," the distinction between them being determined to an extent by the time span encompassed by the model calculation. Some slowly reacting compounds are shown as examples in the boxes in Figure 3.1. The fate of nonreactive materials, insofar as they influence the oceans, is through an absorption or "dry deposition" process to the sea surface. The gaseous pollutants that are more reactive, soluble, or both, such as $SO_2$, typically undergo chemical reactions to form fine particles, with diameters less than 1 $\mu$m. These particles are, thus, secondary pollutants that move into the particulate phase of the model. These same reactive species can also be attached to the surface of the existing aerosols and then be absorbed by the aerosols. (Specific chemical processes are discussed in later chapters; some detailed mechanisms are presented in Chapter 8.)

Particles larger than 1 $\mu$m are usually generated directly from soils, from the bursting of bubbles at sea, and from certain anthropogenic combustion processes (fly ash, for example). Accordingly, in Figure 3.1, particles are subdivided into small (or submicrometer) and large (or supermicrometer) size ranges. The latter particles may be either soluble or insoluble; in most cases, they are removed rather rapidly as a consequence of gravitational sedimentation, precipitation scavenging, or both. However, large (>10 $\mu$m diameter) soil particles have been observed on ships, aircraft, and islands in midoceans (Carlson and Prospero, 1972; Prospero and Nees, 1977) at distances of thousands of kilometers from the sources. Thus, under certain atmospheric conditions, both small and large particles can participate in long-range transport despite differences in the sedimentation velocity.

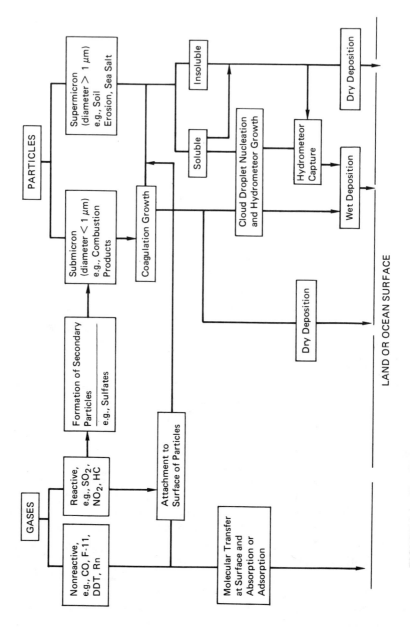

FIGURE 3.1 Simplified schematic of pollutant types and transformations during transport to oceans.

Figure 3.1 is intended to serve as a useful reminder of the minimum number of transformation processes that need to be taken into account when modeling a given type of pollutant. Because of the variability of sources and sinks, the complexity of chemical and surface reactions, and the lack of any substantial worldwide data set for the concentration of airborne materials, the development of a model as depicted in the figure is no simple matter.

## III. MODEL TYPES

Transport models of atmospheric flow can be classified into at least four types, dependent on the methods and techniques used to determine the three-dimensional transporting wind field or the effects thereof:

1. Diagnostic models in which direct observations of the atmospheric wind and mass fields are utilized for the calculation of the transport and dispersion of trace constituents;

2. Numerical weather-prediction models in which the transporting wind and mass field is obtained by a prediction or forecast from a numerical model;

3. General circulation models (GCM's) in which self-consistent circulations are generated from the governing physical equations;

4. Two-dimensional, zonal-height, mean circulation, and diffusion models.

### A. DIAGNOSTIC MODELS

These are useful for diagnosing events as they occurred in an actual situation. Because it is based on observed data, the accuracy of the transport prediction depends on the distribution of observations rather than on the representativeness of a numerical prediction method.

Radiosonde measurements of temperature, pressure, relative humidity, and the horizontal velocity can be utilized to compute the three-dimensional trajectories of the center of mass of a trace constituent and the dispersion relative to the moving center of mass. Errors in these computations are due primarily to the large time interval between observations (12 h), the variable distances between the radiosonde stations (100 to >1000 km), the errors in the reported wind velocities, and the inability to measure the small, but very important, vertical velocities.

To determine the trajectory of the center of mass involves a solution of the following equation for successive 12-h intervals:

$$\mathbf{r}_2(t + 12 \text{ h}) - \mathbf{r}_1(t) + \int_t^{t+12\text{h}} \mathbf{V}(x,y,z,t)\,dt, \tag{3.2}$$

where $\mathbf{r}$ is the position vector and $\mathbf{V}$ is the representative mean velocity for the center of mass. Because $\mathbf{V}$ varies in space and time, it is necessary first to generate continuous fields from the discrete array of radiosonde observations. Objective methods for generating continuous spatial distributions from the observed winds are available (Eddy, 1967; Danielsen and Deaven, 1974). In general, these analyses are more reliable north of 20° N latitude than they are between 20° N and the equator. North of 20° N, the errors in $\mathbf{V}$ range from 0 to 10 m/sec with the average error about 2–3 m/sec.

Having derived "representative" winds at gridpoints in $x,y,z$ space, objective, nonlinear interpolations can be used to obtain values at any point. But direct measurements of the vertical velocities are not available. To obtain reasonable approximations to the net 12-h vertical displacements, one can take advantage of the quasi-isentropic nature of flow in the troposphere by determining 12-h trajectories on isentropic surfaces.

When the relative humidity (RH) is less than 100 percent, a surface of constant entropy is also a surface of constant potential–virtual temperature. The virtual temperature is defined by

$$T_v = T(1 + 0.621\chi_v), \tag{3.3}$$

where $T$ is the actual temperature and $\chi_v$ is the water-vapor-to-air mass mixing ratio. Water molecules, being lighter than air molecules, decrease the mean density of the mixture. If the gas constant for dry air is used, then the temperature must be augmented to account, in the equation of state, for the decrease in density.

The potential–virtual temperature, $\theta_v$, is given by

$$\theta_v = T_v \left(\frac{1000}{p}\right)^{0.286}, \tag{3.4}$$

where the pressure $p$ is expressed in millibars. The values of $\theta_v$ are not affected by compressional heating and are thus analogous to the temperature in a liquid. When air descends and compresses, $T_v$ increases, but little or no energy is lost by conduction or radiation; therefore, $\theta_v$ essentially remains constant.

The major advantage of analyzing the atmosphere on surfaces of constant entropy (i.e., constant $\theta_v$) is the reduction in trajectory determinations from a three-dimensional problem to a two-dimensional one. The initial and final heights are determined by simply locating the initial and final $x,y$ positions on the $\theta_v$ surface whose height contours are known. The change in height contours over a 12-h time period yields the 12-h mean vertical motion. In other words, one can determine the three-dimensional motion of the center of mass of the system by computing successive 12-h positions on the $\theta_v$ surfaces.

Of course, one knows that the atmosphere is, effectively, a thin, fluid layer surrounding the earth, a layer whose horizontal dimensions are very large compared with its vertical dimensions. For example, the circumference of the earth is 40,000 km, but the tropospheric depth varies from only 8 to 18 km (pole to equator). A representative ratio of $\Delta x/\Delta z$ or $\Delta y/\Delta z$ is $10^3/1$. However, the velocities are similarly scaled, i.e., $u/w \simeq v/w \simeq 10^3/1$; thus, an air parcel can span the vertical depth of the troposphere in about the same time as is required to span a horizontal dimension.

Also, the advective products are approximately independent of direction because the vertical gradients are $\sim 10^3$ times as large as the horizontal gradients; therefore,

$$w \frac{\partial \chi}{\partial z} \simeq u \frac{\partial \chi}{\partial x} \simeq v \frac{\partial \chi}{\partial y} . \tag{3.5}$$

If one considers the anisotropy in both the velocities and the gradients, one concludes that vertical velocities of 1 to 10 cm sec$^{-1}$ (velocities too small to measure by conventional methods) are as effective in transporting trace constituents as are horizontal velocities of 10 to 100 m sec$^{-1}$.

Consistent with this conclusion, one expects, and one observes, that real three-dimensional trajectories are systematically different from those inferred from conventional constant-pressure charts. The air does not flow at constant pressure; it tends to ascend in regions of warm advection and descend in regions of cold advection. The correlation between vertical velocity and the horizontal advection of temperature makes the curvatures of the actual, and of the isentropic, trajectories more anticyclonic than an isobaric trajectory, as shown in Figure 3.2.

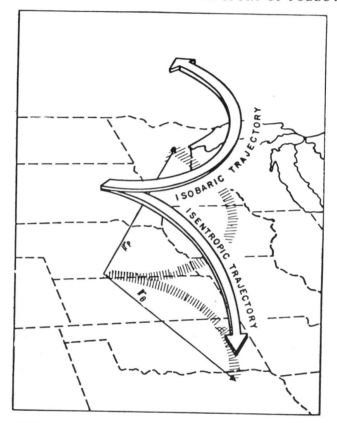

FIGURE 3.2  A descending anticyclonic trajectory (isentropic) compared with a cyclonic trajectory (isobaric) in which the vertical velocities are neglected. Source: Danielsen (1961).

This difference in curvatures can be expressed by

$$K_\theta = K_p + \frac{\partial \gamma}{\partial p} \frac{\omega}{v} \simeq K_p - C \frac{\omega_2}{v_2} , \qquad (3.6)$$

where $K_\theta$ = curvature of the trajectory on a constant $\theta$ surface, $K_p$ = curvature of the trajectory on a constant $p$ surface, $\partial \gamma / \partial p$ is the change in wind direction with pressure ($-z$ direction); $\omega = dp/dt \simeq -\rho g w$ is the rate of change in pressure due to a vertical velocity $w$, and $C$ is a proportionality constant that relates the turning of the wind with temperature advection to $\omega/v$. The difference is very large when $\omega$ is large, i.e., in the baroclinic regions where cyclones are forming, propagating, or both.

The predominance of anticyclonically curved streamlines in the ascending flow to the east of a developing cyclone is shown in Figure 3.3. Air from the boundary layer north of the warm front is rapidly ascending into the upper troposphere, turning anticyclonically and increasing in speed as it forms the tropospheric portion of the main jet stream aloft. (For a review of trajectory methods, see Danielsen, 1974.)

Many three-dimensional trajectories have been computed on isentropic surfaces during large-scale cyclogenesis using the method described by Danielsen (1961). A systematic pattern is evident and essentially repeated in each case study. As the cyclone develops, warm, moist air in advance of the cyclone ascends, as shown in Figure 3.3. Then it travels rapidly eastward in the upper troposphere until it approaches the next downstream wave. Turning anticyclonically, it begins to descend to the west of the wave and fans out in a strong deformation field. The right branch of the flow continues to descend and turn anticyclonically as it approaches, and is mixed down into the boundary layer. The left branch turns cyclonically, reaching its lowest elevation as it passes the wave trough, then begins to ascend again. Strong vertical mixing in this branch occurs when the air overtakes, and is entrained into, a line of cumulonimbus clouds, which form along the cold front in advance of the wave trough.

This long-range transport pattern, shown schematically in Figure 3.4, explains how pollutants or natural aerosols (in this case, soil dust) generated over a continent to the *west* of a major ocean can, in two or three days, reach the ocean surface, in a region dominated by an *easterly* flow. For example, quartz grains raised in Asiatic dust storms can be deposited on the Hawaiian Islands, which are dominated by the northeast tradewinds (Jackson *et al.*, 1973), by the subsiding northwest winds behind a Pacific cold front.

It is important to recognize that only a fraction of the airborne pollutants can survive in the ascending air to participate in long-range transport. During ascent, the air cools by expansion, generating clouds and precipitation. These growing and falling hydrometeors scavenge aerosols, transporting them rapidly downward toward the earth's surface. Some reach the surface; others are resuspended in the air when the hydrometeors evaporate during descent. Only those small cloud droplets and ice crystals whose terminal speeds are negligible will be carried upward with the air to form cirrus clouds or cirrus anvils in the upper troposphere. When these ice particles evaporate, the mineral

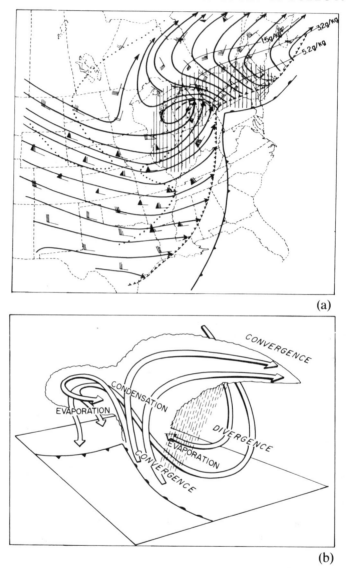

(a)

(b)

FIGURE 3.3   (a) Observed winds and streamlines on a moist isentropic surface. East of the developing wave cyclone the surface slopes upward from the ground (southern New York State) to the tropopause (Hudson's Bay, Canada). Saturated specific humidities are expressed in grams of water vapor per kilogram of air. Wind speeds in knots are denoted by barbs to the left of the vector. Each barb equals 10 knots; a triangle equals 50 knots. The cross-hatched area denotes a region of observed precipitation. (b) Schematic perspective drawing of predominantly anticyclonic flow in air ascending from and descending toward the boundary layer. Source: Danielsen and Bleck (1967).

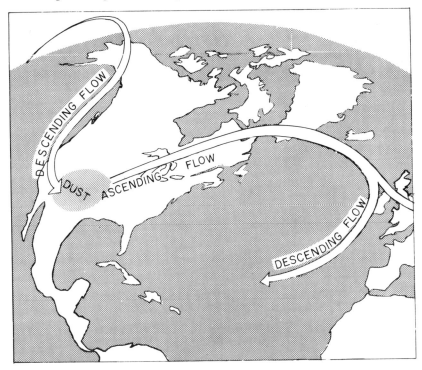

FIGURE 3.4  Schematic of long-range transport (three-dimensional) from major dust storm to Atlantic Ocean. Source: Jackson *et al.* (1973).

and salt residues remain aloft as new aerosols that can participate in long-range transport.

Thus, some pollutants, lifted from the boundary layer by cyclonic storms over the central and eastern United States, will be returned to the surface of the continent in precipitation before the air flows over the Atlantic Ocean, but some will be carried aloft, eastward, to the midportions and eastern portions of the Atlantic Ocean.

Another type of diagnostic model, useful for modeling transport in the planetary boundary layer, is the Gaussian plume, or puff, model (Reiter, 1976; Heffter, 1975). Usually the transport portion of this type of model uses observed winds to calculate mean horizontal trajectories, which are determined by averaging winds through an arbitrarily specified vertical layer.

In the diffusion section of Reiter's (1976) model, vertical mixing is assumed to occur instantaneously up to the top of the mixed layer. The latter's height is judged from inspection of observed wind profiles rather than being related to the thermal structure. Vertical transport by convective clouds is not yet treated. Lateral mixing, relative to the Lagrangian mean trajectory, is taken to yield a Gaussian distribution of pollutants with standard deviation, $\sigma_y$, given by $\sigma_y = 0.014 v_g t$, where $v_g$ is the geostrophic wind speed and $t$ is the elapsed time following pollutant release.

Heffter's (1975) model employs a different parameterization for the vertical spread. Here, the vertical distribution of pollutant concentration is assumed to be Gaussian with a standard deviation, $\sigma_z$, proportional to $(K_z t)^{1/2}$, where $K_z$ is a constant vertical eddy diffusivity. The net effect of a variable boundary-layer height and intermediate cloud transports are thus lumped together into a single time-dependent $\sigma_z$. Lateral diffusion across the trajectory is assumed given by $\sigma_y = \sigma_v t$, with $\sigma_v$ set to 0.5 m sec$^{-1}$.

## B. NUMERICAL WEATHER-PREDICTION MODELS

Transport models of this second type are potentially useful for forecasting the future state of a trace constituent for the period of the forecast. Such models have been used for Lagrangian trajectory forecast (Reap, 1972) and Eulerian field predictions (Kreitzberg et al., 1977). The major disadvantage of this approach is that the results are limited by the progressive degradation of the forecast beyond a few days. In addition, the forecasted large-scale winds, vertical motions, and mass fields are necessarily oversmoothed because of computational stability constraints of the numerical models.

The chief advantage of this approach is that the transport model can make use of the wind and mass fields generated by the numerical model, with a major part of the computer storage remaining for parameterization and modeling of the chemical interactions and reactions. Because the same transport winds are utilized for each simulation, the role of the chemistry can be investigated by varying parameters and chemical modeling techniques during each simulation.

Another mode of operation for models of this type is to update the numerical forecast with actual synoptic data every 12 h. However, numerical weather-prediction models suffer from initialization shocks produced by imbalances between the initial momentum and mass fields. Hence, these models do not accurately simulate the atmosphere for the first few hours of the forecast periods and are less reliable during this initial period.

C. GENERAL CIRCULATION MODELS

The third modeling possibility is to utilize winds as generated in the so-called general circulation models (Hunt and Manabe, 1968; Mahlman, 1973; Cunnold *et al.*, 1975). These models simulate self-consistent circulations from the governing physical equations. GCM's are very useful for determining longer-term behavior of trace constituents, provided that they are capable of properly simulating local meteorological structure and the longer-term statistical behavior of the atmosphere for the important space and time scales. A disadvantage of this approach is that such models are not developed for the express purpose of simulating particular weather situations. Instead, the models simulate their own sequence of weather events, starting from initial data having a climatology similar to that of the real atmosphere but having local weather patterns that are different. With regard to the pollutant transport problem, specific events cannot be predicted, but long-term patterns could be.

In a comprehensive model incorporating trace-constituent behavior, the various physical, chemical, and transport processes rarely operate independently of one another. In spite of this, it is necessary that the description of each individual process or mechanism be as realistic as possible when viewed as a separate entity. This increases assurance that the system is working meaningfully when all component parts of the model are brought together.

The most obvious requirement for accurate modeling of the effects of large-scale wind transport is that the advecting wind field be realistic in terms of its physical and statistical characteristics. This problem is especially difficult but has received major attention for the past 20 years (Phillips, 1956; Smagorinsky, 1963; Manabe *et al.*, 1965; Mintz, 1968; Kasahara and Washington, 1967; Miyakoda *et al.*, 1969; Holloway and Manabe, 1971; Manabe and Holloway, 1975). The parallel increases in our fundamental understanding of physical processes and in the available computational power have led to more realistic atmospheric simulations. However, problems do remain for a large variety of reasons.

For example, even if the large-scale wind field is accurately simulated, it is still possible that large inaccuracies can occur in the simulated trace-constituent behavior. The most direct difficulty results from the imperfections of various numerical transporting algorithms due to truncation error (Crowley, 1968; Orszag, 1971; Mahlman and Sinclair, 1977.)

An indirect, but related, difficulty can occur in GCM's even when advection is allowed to dominate diffusive processes. As elements of a tracer become sheared and stretched by the advecting wind field, the tracer gradients become so large that the associated increase of truncation error leads to situations in which the calculated mixing ratio becomes negative. This normally requires a preventative response in the model, either a mass-conserving "repair" of the negative value (Mahlman, 1973) or a subgrid-scale transfer process, which acts to pre-empt the occurrence of the negative value.

In any numerical model, the finite capabilities of the computer system require that the atmosphere be approximately represented by a finite resolution in space and time, necessitating that the transfer between the large-scale and the subgrid scale be parameterized in terms of the resolved large-scale structure. At present, there is no fully satisfactory, or universally accepted, way of accomplishing this because the separation between the resolved scales and the important unresolved scales is so large. Much of the difficulty arises because the parameterized subgrid scale transport must take a number of nearly unrelated processes into account, e.g., dry and moist cloud-scale convection, clear-air turbulence, mesoscale disturbances, orographic effects, nonlinear scale transfers in three-dimensions, stretching and shearing of tracer elements by large-scale winds into subgrid-scale dimensions, small-scale turbulence, and boundary-layer effects.

An example of the type of parameterization that is required for vertical and horizontal mixing in general circulation models is described by Smagorinsky et al. (1965). In this model, the vertical diffusivity is assumed to exist only in the lowest few layers. Its value is taken as $K_z = L^2/\partial V/\partial z$, where $L$ is a mixing length prescribed to be 27 m in the lowest layer of the model, decreasing to zero above a height of the order of 1 km. This formulation of diffusivity has been used both in the prediction equation for winds and temperature and for the tracer substance or pollutant. However, it does not take into account the enhanced vertical mixing that occurs in the presence of buoyantly driven turbulence when the vertical wind shear is at a minimum.

Lateral mixing of a tracer, $\psi$, on a scale smaller than the GCM grid length, $\Delta x$, is assumed to be governed by eddy diffusion with a horizontal eddy diffusivity given by

$$K_x \propto (\Delta x)^2 D \left( \frac{\partial ln\bar{\psi}}{\partial x} \Delta x \right)^2 \qquad (3.7)$$

for the $x$ component of the mixing, where $D$ is the magnitude of the velocity deformation.

Tracer removal, as simulated by numerical models, is almost always closely related to the processes mentioned previously. The most obvious of these are local tracer removal by large-scale and convective precipitation, chemical destruction, and dry removal at the surface. These processes must be parameterized specifically in terms of the resolved scale variables and the parameterized subgrid-scale structure.

Any comprehensive calculation of tropospheric trace-constituent behavior must incorporate all the above-mentioned processes into the model framework. This demands that the processes not only be physically realistic when viewed individually but also computationally compatible with other components of the model.

## D. TWO-DIMENSIONAL TRANSPORT MODELS

Two-dimensional transport models are most useful for predicting the large-scale, long-period distributions of trace constituents. To date, these models simulate the zonal–seasonal mean concentrations and depositions in a meridional–vertical plane. Transport by both mean and eddy motions is included in these models. Transport by mean meridional circulations is explicit; values of $\bar{v}$ and $\bar{w}$ are specified at each grid point. Transport by eddies is usually parameterized as the scalar product of a diffusion tensor and the gradient of the mean concentration or mean mixing ratio of the tracer.

A diffusion tensor is required because it must simulate the diffusive effects of the complete spectrum of eddies and internal waves. The latter, particularly the largest-scale waves, dominate the diffusion, and their motions are definitely anisotropic. In the midtroposphere, their principal axis of diffusion is inclined upward toward the north at approximately half of the slope of the mean isentropic surfaces. In the lower stratosphere, their principal axis is inclined downward to the north at a slope that exceeds the slope of the mean isentropes.

Models based only on diffusion have been developed by Reed and German (1965) and by Davidson *et al.* (1966), but in general both the mean and turbulent transports must be included. As shown by Louis and Danielsen (1976), these two transports act to oppose each other; therefore, their residual controls the effective transport.

The main advantages of a two-dimensional model are the elimination of the third spatial dimension and the extention of the time step. The tremendous reduction in computation time and in storage requirements enables one to include more detailed photochemical and chemical-physical processes. The major disadvantages are the present uncertainties in the magnitudes and distributions of the mean velocities and the components of the diffusion tensor. Our inability to measure directly the vertical velocity of the air prevents an objective evaluation of both the mean and turbulent components. In general, some subjective assumptions must be used to derive the components, and, therefore, *every model must be tuned by comparisons between the predicted and observed distributions of one or more tracers.*

To study the transport of continental trace gases and aerosols to the oceans, it would also be helpful to develop a mean longitude–height model by averaging over all latitudes and time. In such a model, the mean continents and oceans (in latitudinally averaged bands) could be simulated, and the longitudinal variation of transport to the oceans could be predicted.

## IV. DEVELOPMENTS IN PARAMETERIZATION OF BOUNDARY-LAYER DIFFUSION

The success of any large-scale transport model will be critically dependent on the success with which the boundary-layer processes are modeled. A mechanistic approach currently under development for use within a GCM involves the prediction of the boundary layer (BL) height, $h$, from a rate equation. Surface fluxes of heat and momentum support turbulence, which tends to drive $h$ upward through the entrainment mechanism (e.g., see Tennekes, 1973; Carson, 1973; Deardorff, 1974); stable stratification along with large-scale subsidence and cloud-induced subsidence all tend to drive $h$ downward. A quasi-steady state usually obtains, with $h$ being of the order of 500 m over the oceans and, over land, of order 100 m at night and 200 m to 3 km in the daytime. Within this BL, the pollutant could be treated as well mixed vertically, with value $C_{im}$. Its value may be predicted at each grid point of the model from Eq. (3.1) upon setting the vertical mixing term to

$$\frac{\partial}{\partial z} (\overline{w'C_i'})_m = (\overline{w'C_i'})_l - (\overline{w'C_i'})_0/h, \tag{3.8}$$

where the subscript $m$ stands for a mixed-layer mean. The vertical flux $(-\overline{w'C_i'})_0$ at the surface is taken as the deposition rate in the absence of pollutant sources, and the vertical flux $(\overline{w'C_i'})_l$ refers to the flux at a level just below $z = h$ given by

$$(\overline{w'C_i'})_l = -w_e(C_{i2}-C_{im}), \tag{3.9}$$

where $w_e$ is the entrainment rate ($w_e \equiv dh/dt - \overline{w}_h$), already utilized in predicting $h$, and $C_{i2}$ is the mean value of $C_i$ estimated (from inspection of $C$ at higher levels within the model) to exist just above the top of the mixed layer. In the case of a polluted mixed layer rising into a clear air mass, $C_{i2} = 0$. In the case of the stable BL, as at night over land, the entrainment flux $\overline{(w'C_1')}_1$ may be neglected.

The lifting condensation level (LCL) is also predicted in this parameterization, with BL clouds being of the stratocumulus variety when LCL $< h$, provided the top of the mixed layer is capped by a sufficiently strong inversion. If LCL $>> h$, no BL clouds appear at all, and for intermediate cases scattered cumulus clouds of fractional area $a$ are predicted from an interpolation formula that is a function of LCL/$h$. Cloud-induced subsidence is currently set proportional to $a(1 - a)$ but should later be made a function of the cloud-mass flux deduced from a subgrid-scale precipitation parameterization (Betts, 1973). A reasonable cloud parameterization will also involve a mean-cloud-depth scale, which, when utilized with the cloud-mass flux, would permit a prediction of the vertical pollutant flux associated with cloud transports. Although deep clouds, which precipitate heavily, act as strong sinks for particles, their upward pollutant transport is still important because of the huge export of cloud mass (probably polluted) that can occur in cloud anvils. Cloud transports are of even greater importance for the nonreactive gaseous pollutants or tracers not subjected to precipitation scavenging. Also, cumuli of intermediate heights that do not reach the precipitation stage still serve as effective polluters of the cloud layer, while helping to dilute the subcloud or mixed layer.

Although the development of this kind of boundary-layer parameterization is a time-consuming task, the additional effort needed to include the pollutant conservation equation is relatively minor for a *non*reactive gas.

A comprehensive boundary-layer parameterization for a diagnostic-trajectory pollution transport model could also be developed, which would make use of observed precipitation regions and observed boundary-layer heights.

## V.  INHERENT COMPLEXITIES IN POLLUTION-TRANSPORT MODELING

Pollution-transport modeling is in its infancy for fundamental reasons. A reliable transport model must be developed and tested before the additional complexities of the gases and aerosols can be included.

These models have progressed from simple, one-level barotropic models, applicable at midtropospheric levels, to multilevel baroclinic models, applicable to the complete troposphere and most of the stratosphere. The effects of topography, solar heating, and the water-vapor cycle are now being included, but the latter is generally oversimplified by parameterizations.

The water-vapor cycle is, of course, directly influenced by gases and aerosols, and vice versa. Many chemical reactions depend on the relative, or absolute, humidity. Also, the aerosols acting as condensation and freezing nuclei affect the precipitation type, intensity, and amount. The transport of even a nonreactive pollutant species, therefore, requires a rather comprehensive meteorological model, which includes the effects—explicit or parameterized—of the water-vapor cycle.

For these reasons, it can be stated that *it is highly unlikely that a comprehensive model, capable of predicting the transport of reactive gases and aerosols, will be available in the near future.* In order to determine the most significant reactions and to develop simpler parameters to be used in the transport models, it will be necessary to create detailed microphysical, photochemical, and cloud models; to develop new sensors (remote and *in situ*); and to carry out extensive laboratory and field experiments.

Listed below are specific subject areas requiring further research; these are grouped according to substance and process.

*Gases*

Homogeneous reactions

(a) Source strengths of primary pollutants (spatial–temporal distribution)
(b) Concentrations of transient species
(c) Ultraviolet light intensity
(d) Absolute humidity measurements for complete range of atmospheric temperatures
(e) Reaction rates over a range of pressures, temperature, and concentrations

Heterogeneous reactions

(f) Surface reaction rates for a variety of particles
(g) Solubility of gases in solution particles

*Particles*

(a) Source strengths for combustion processes and for the wind generation of particles from soils and the sea
(b) Vertical, areal, and temporal statistics on the concentration, size distribution, and chemical composition of primary and secondary aerosols

*Removal Processes*

(a) Nucleating ability of aerosols (condensation and freezing)
(b) Collision and collection efficiencies (aerosol–hydrometeor, aerosol–aerosol)
(c) Dry deposition rates of gases and aerosols
(d) Solubility of gases in ocean

In addition to these requirements, more measurements are needed to determine the subgrid-scale diffusion appropriate to any given model and the significant feedbacks from gases and particles (primarily through interactions with water vapor and radiation) to the thermodynamics and dynamics of the model.

## VI. SUMMARY: MODEL APPLICATION AND VALIDATION

Because of the numerous uncertainties listed above, pollution-transport models will require validation. At present, a variety of transport models are available that might be applicable for the transport of specific pollutants from continents to ocean; but none of these is a comprehensive, pollution-transport model. The range of applicability and the limitations of these models are summarized in Table 3.1.

The *diagnostic models* can be used to trace the three-dimensional trajectory of the center of mass and the large-scale dispersion of a pollutant, either forward in time from a source to a receptor site (a sensor or a deposition area) or backward in time from a receptor site to a source. Validation is usually based on observations made aboard aircraft carrying instruments that can measure a pollutant species continuously and at low concentrations. In the past, radioactive particles from stratospheric nuclear tests and ozone from the lower stratosphere have been used successfully as tracers to validate isentropic trajectory models. The large-scale deformations often concentrate the

TABLE 3.1 Summary of Transport Models

| Type | Independent Variables | Transport by | Applicability | Time Scale of Sampling |
|---|---|---|---|---|
| Diagnostic methods | $x,y,\theta_v,t$ | Observed $u,v$ winds at constant $\theta_v$ | Trace transport from explicit source for periods of a few days or determine source for an explicit pollution event | Continuous high-resolution sampling preferred. If integrated samples are unavoidable, integration time should be <1 h |
| Predictive models | $x,y,z,t$ or $u,y,p,t$ | Predicted $u,v,w$, or $u,v,\omega$ components and subgrid-scale diffusion fields periodically adjusted by observed data | Predict transport for a few days to a week | Sampling integration time <6 h |
| General circulation models | $x,y,z,t$ or $x,y,p,t$ | Predicted $u,v,w$ or $u,v,\omega$ components and subgrid-scale diffusion | Predicted statistically representative geographical distributions for a season or a year | Sampling integration times of a few days to a week are acceptable |
| Two-dimensional models | $\theta,z,\tau$ or $\lambda,z,\tau$ | Specified $\bar{v},\bar{w}$ mean winds and parameterized eddy diffusion or specified $\bar{u},\bar{w}$ mean winds, etc. | Predict meridional distribution of zonal-seasonal means for several years  Predict longitudinal distribution of meridional-seasonal means, etc. | Sampling integration times of the order of months or seasons  Sampling integration times of the order of months or seasons |

46

tracer into thin laminae; consequently, continuous (rather than integrated discrete) measurements are preferred for validation. If the sampling integration time exceeds a few hours, then the data cannot be used to validate the diagnostic model.

The *numerical weather-prediction model* is best suited for predicting pollution transport for periods of the order of days where high temporal resolution is not required. The combined effects of phase errors in the predicted wind fields and truncation errors, which tend to underpredict the transport speeds in the jets, cause phase errors in the predicted transport. The grid distance of about 350 km also imposes an effective 5–10 h filter on the predictions; consequently, pollutant or tracer measurements can be integrated over periods up to about 6 h. Of course, if the grid distance is reduced, the equivalent averaging time should also be reduced.

The *general circulation models* are most appropriately applied to long-term averages, weekly or monthly means. Although the resolution is comparable with that of the weather-prediction models, the GCM is not initiated, or periodically forced, by observed data. Therefore, the storms predicted by the GCM will resemble actual storms, but it is purely coincidental if they are in the right place at the right time. These models should be useful for predicting the global tropospheric distributions of those pollutants that have continuous sources on continents and for predicting the mean fluxes to the ocean surface.

The *two-dimensional mean models* are obviously suited to long-range predictions and long-term averages. One advantage of these models is the small amount of computer time required for the dynamics. Consequently, time is available to include more detailed photochemical and microphysical computations. Models of this type are now being used to test and reduce the complex set of chemical reactions to a dominant subset that is appropriate for the chemically active stratosphere and the lower troposphere.

## VII.  CONCLUSIONS AND RECOMMENDATIONS

Transport modeling is now developing at an accelerating rate so that validating experiments are becoming both possible and necessary. Based on an assessment of current capabilities, the panel recommends the following:

1. *An experiment should be performed in some coastal urban region to measure the direct transport within the boundary layer to the ocean.*

Here, the problem is reduced to one of quasi-two-dimensional transport. Aircraft would be required to measure the meteorological parameters and the vertical and areal concentration fields of aerosols and gases both inside the boundary layer and for a short distance above it. Ships would be required to measure concentrations in surface level air and also to measure the flux across the air–sea interface. A coastal transport study has a number of advantages over one carried out over the continent—the topography is uniformly flat, and there are no downwind sources to complicate the input-term computations. On the basis of meteorological considerations and the distribution of major sources, the most logical site for such an experiment would be in the region off the northeast coast of the United States.

2. *Programs should be developed to utilize certain materials that are currently being emitted into the atmosphere as tracers for large-scale transport model validation.* Natural and anthropogenic processes generate a number of substances that could be useful as tracers. Examples are the fluorocarbons (see Chapter 6), carbon monoxide and carbon dioxide (see Chapter 8), and $^{222}$Rn (see Chapter 9). These materials are gaseous and relatively inert, and they are emitted exclusively, or primarily, from the continents. Radon-222 is entirely natural in origin, while the fluorocarbons are entirely anthropogenic; the oxides of carbon are derived from both sources. There are a number of programs concerned with the study of these and other materials in the environment. The data output from such programs *potentially* could be useful for validating transport models. *However, in many programs, including most of the routine monitoring programs, the frequency of sampling and the duration of individual sampling periods is such as to preclude the use of the resulting data for model-validation purposes.* Consideration should be given to the feasibility of modifying these programs so that the data could be used for model testing. One approach may be to establish specific time periods, worldwide, when programs could switch to a suitable data-gathering mode.

3. *The modeling of the transport of chemically reactive materials will require detailed knowledge of the concentration and distribution of the dominant chemical species in the atmosphere and of the controlling chemical reactions and physical-chemical processes.* This can only be done by simultaneously measuring aerosols and a wide variety of gases in the atmosphere both within, and above, the marine boundary layer using highly instrumented aircraft. Such experiments will require the knowledge, technical capabilities, and close cooperation of scientists from universities, government laboratories, and research centers. Experiments of this nature are feasible, and some are presently being planned. *Such cooperative efforts must be strongly encouraged.*

4. Many large-scale, international meteorological programs are currently being planned, and some, such as the First GARP Global Experiment, are close to execution. In many cases, these experiments will take place in regions where there is a dearth of data on the concentrations of important chemical species in the atmosphere and in the oceans. *Chemists should be encouraged to participate in these meteorological field experiments.*

5. *Within five to ten years, it should be possible to conduct a major experiment to test modeling concepts of long-range transport from the continents to the ocean.* This type of transport, being three-dimensional and involving, among other processes, cloud formation and subsequent hydrometeor removal of aerosols and soluble gases, will be most difficult to monitor. A multiaircraft experiment will be essential, and a comprehensive array of surface monitors (ships and instrumented buoys) will also be required. The positions of the ships and the flight schedules and paths must be determined in advance with suitable lead times; this will require a good capability for fast and accurate predictions of the development, growth, and movement of specific meteorological phenomena such as a storm or a front. An experiment of this type should be international in scope, involving chemists, particle and cloud physicists, meteorologists, and oceanographers. *To plan such an experiment now may be premature, but the time is right to discuss its possibility and to engender interest.*

## APPENDIX 3.A   SYMBOLS

$p$, pressure
$\rho$, air density
$T$, temperature
$T_v$, virtual temperature
$\theta_v$, potential virtual temperature
$\chi_v$, water-vapor-to-air mixing ratio
$\mathbf{r}$, position vector
$\mathbf{u},\mathbf{v},\mathbf{w}$, velocity vector components
$\nabla$, gradient (del) operator
$V$, horizontal wind speed
$V_g$, geostrophic wind speed
$\gamma$, wind direction
$\omega$, $dp/dt$, vertical "velocity" in pressure coordinate system
$K_\theta$, curvature of constant $\theta$ trajectory
$K_p$, curvature of constant $p$ trajectory
$K_z$, vertical eddy diffusivity

$K_{ij}$, reaction rates between species $i$ and $j$
$C_i$, concentration of species $i$
$C$, constant
$g$, acceleration due to gravity
$(\overline{\phantom{x}})$, space–time mean
$(\ )'$, $(\ ) - (\overline{\phantom{x}})$, deviation from mean
$\sigma y_1 z$, standard deviation in $y, z$ direction
$L$, mixing length
$R_n$, radon
$\tau_{ri}$, radioactive half-life of species $i$
$\chi, \psi$, mixing ratio of a tracer
BL, boundary layer
$h$, height of boundary layer
$W_e$, boundary-layer entrainment rate
LCL, lifting condensation level
$D$, magnitude of velocity deformation

## REFERENCES

Betts, A. K. (1973). Non-precipitating cumulus convection and its parameterization, *Quart. J. R. Meteorol. Soc. 99*, 178–196.

Carlson, T. N., and J. M. Prospero (1972). The large scale movement of Saharan air outbreaks over the equatorial North Atlantic, *J. Appl. Meteorol. 11*, 283–297.

Carson, D. J. (1973). The development of a dry inversion-capped convectively unstable boundary layer, *Quart. J. R. Meteorol. Soc. 99*, 450.

Crowley, W. P. (1968). Numerical advection experiments, *Monthly Weather Rev. 96*, 562.

Cunnold, D., F. Alyea, N. Phillips, and R. Prinn (1975). A three-dimensional dynamical-chemical model of atmospheric ozone. *J. Atmos. Sci. 32*, 170.

Danielsen, E. F. (1961). Trajectories: Isobaric, isentropic and actual, *J. Meteorol. 18*, 479–486.

Danielsen, E. F. (1974). Review of trajectory methods, *Advan. Geophys. 18B*, 73–94.

Danielsen, E. F., and R. Bleck (1967). Research in four-dimensional diagnosis of cyclonic storm systems, Air Force Cambridge Research Laboratories, AFCRL-67-0617, pp. 1–34.

Danielsen, E. F., and D. G. Deaven (1974). Northern hemispheric analysis methods for deriving two-dimensional transport model of troposphere and stratosphere from isentropic trajectories and potential vorticity, *Proc., IAMAP/IAPSO Combined First Special Assemblies, II*, Melbourne, Australia, pp. 835–848.

Davidson, B., J. P. Friend, and H. Seitz (1966). Numerical models of diffusion and rainout of stratospheric radioactive material, *Tellus 18*, 305–315.

Deardorff, J. W. (1974). Three-dimensional numerical study of the height and mean structure of a heated planetary boundary layer, *Boundary Layer Meteorol. 7*, 81.

Donaldson, C. (1973). Atmospheric turbulence and the dispersal of atmospheric pollutants, EPA-R4-73-016a (Vol. 1).

Eddy, A. (1967). Statistical objective analysis of scalar data fields, *J. Appl. Meteorol. 6*, 597–609.

Heffter, J. L. (1975). A regional–continental scale, transport, diffusion, and deposition model. Part I. Trajectory model; Part II. Diffusion-deposition models. NOAA-TM-ERL-ARL-50, NTIS No. COM-75-11094-0G1.

Hilst, G. R., *et al.* (1973). The development and preliminary application of an invariant coupled diffusion and chemistry model, NASA-CR-2295, 82 pp.

Holloway, J. L., Jr., and S. Manabe (1971). Simulation of climate by a global general circulation model. I. Hydrologic cycle and heat balance, *Mon. Weather Rev. 99*, 335.

Hunt, B. G., and S. Manabe (1968). Experiments with a stratospheric general circulation model. II. Large-scale diffusion of tracers in the stratosphere, *Mon. Weather Rev. 96*, 503.

Jackson, M. L., D. A. Gillette, E. F. Danielsen, I. H. Blifford, R. A. Bryson, and J. K. Syers (1973). Global dustfall during the Quarternary as related to environments, *Soil Sci. 116*, 135–145.

Kasahara, A., and W. M. Washington (1967). NCAR global circulation model of the atmosphere, *Mon. Weather Rev. 95*, 389.

Kreitzberg, C. W., M. Lutz, and D. J. Perkey (1977). Precipitation cleansing in a numerical weather prediction model, in *The Fate of Pollutants in the Air and Water Environments, Part 1*, I. J. Suffet, ed., Advanced Enrivonmental Science and Technology, Vol. 8, pp. 323–351, John Wiley and Sons, New York.

Louis, J. F., and E. F. Danielsen (1976). An internally consistent two-dimensional transport model, unpublished manuscript.

Mahlman, J. D. (1973). Preliminary results from a three-dimensional general-circulation/ tracer model, *Proceedings of the Second Conference on the Climatic Impact Assessment Program* (A. S. Broderick, ed.), DOT-TSC-OST-73-4, p. 321.

Mahlman, J. D., and R. W. Sinclair (1977). Tests of various numerical algorithms applied to a simple trace constituent air transport problem, in *Fate of Pollutants in the Air and Water Environments, Part 1*, I. J. Suffet, ed., Advanced Environmental Science and Technology, Vol. 8, John Wiley and Sons, Inc., New York.

Manabe, S., J. Smagorinsky, and R. F. Strickler (1965). Simulated climatology of a general circulation model with a hydrologic cycle, *Mon. Weather Rev. 93*, 769.

Manabe, S., and J. L. Holloway, Jr. (1975). The seasonal variation of the hydrologic cycle as simulated by a global model of the atmosphere, *J. Geophys. Res. 80*, 1617.

Mintz, Y. (1968). Very long term global integration of the primitive equations of atmospheric motion. An experiment in climate simulation, *Meteorol. Monog. 8*, 20.

Miyakoda, K., J. Smagorinsky, R. F. Strickler, and G. D. Hembree (1969). Experimental extended predictions with a nine-level hemispheric model, *Mon. Weather Rev. 97*, 1–76.

Orszag, S. A. (1971). Numerical simulation of incompressible flows within simple boundaries: accuracy, *J. Fluid Mech. 49*, 75.

Phillips, N. A. (1956). The general circulation of the atmosphere: A numerical experiment, *Quart. J. R. Meteorol. Soc. 82*, 123.

Prospero, J. M., and R. T. Nees (1977). Dust concentration in the atmosphere of the equatorial N. Atlantic: Possible relationship to the Sahelian drought, *Science 196*, 1196–1198.

Reap, R. M. (1972). An operational three-dimensional trajectory model. *J. Appl. Meteorol. 11*, 1193.

Reed, R. J., and K. E. German (1965). Contribution to the problem of stratospheric diffusion by large-scale mixing, *Mon. Weather Rev. 93*, 313–321.

Reiter, E. R. (1976). *Radioactive Tracers, Part 4, Atmospheric Transport Processes,* ERDA, Critical Review Series.

Smagorinsky, J. (1963). General circulation experiments with the primitive equations: I. The basic experiment, *Mon. Weather Rev. 91*, 99.

Smagorinsky, J., S. Manabe, and J. L. Holloway, Jr. (1965). Numerical results from a nine-level general circulation model of the atmosphere, *Mon. Weather Rev. 93*, 727–768.

Tennekes, H. (1973). A model for the dynamics of the inversion above a convective boundary layer, *J. Atmos. Sci. 30*, 558.

# 4 Wet and Dry Removal Processes

## I. INTRODUCTION

The transport of atmospheric trace constituents through the air–sea interface is the subject of this chapter. The governing transport processes are generally referred to as precipitation scavenging, dry deposition, and resuspension. The objectives of this report are to review some of the known aspects of these processes, identify unknowns, and suggest topics for future research. Although other removal mechanisms are important for some atmospheric trace constituents, e.g., chemical transformations of reactive gases, the scope of the chapter is purposefully restricted to focus on the wet and dry removal processes, especially those aspects relevant to describing air-pollution fluxes to the ocean. A summary of research topics recommended for consideration is presented in the final section of the chapter.

## II. CONTINUITY EQUATIONS AND PARAMETERIZATIONS

A framework for the entire presentation is provided by coupled continuity equations that describe the instantaneous concentrations of the

Members of the Working Group on Wet and Dry Removal Processes were W. G. N. Slinn, *chairman*; L. Hasse, B. B. Hicks, A. W. Hogan, D. Lal, P. S. Liss, K. O. Munnich, G. A. Sehmel, and O. Vittori.

trace constituent in air, $\chi$, and in the ocean, $C$. In generally standard notation (also see Appendix to this chapter) these convective–diffusion equations are

$$\frac{\partial \chi}{\partial t} + \mathbf{v}_g \cdot \nabla \chi = -\nabla \cdot (-D_g \nabla \chi + \chi \mathbf{d}_g) + G_g(\chi) - L_g(\chi), \qquad (4.1)$$

$$\frac{\partial C}{\partial t} + \mathbf{v}_l \cdot \nabla C = -\nabla \cdot (-D_l \nabla C + C \mathbf{d}_l) + G_l(C) - L_l(C). \qquad (4.2)$$

In Eqs. (4.1) and (4.2) the subscripts $g$ and $l$ refer to conditions in the gas phase (i.e., in the atmosphere) and in the liquid phase (ocean), respectively; the $\mathbf{v}$'s are instantaneous (molecular averaged) velocities of the two fluids (assumed incompressible); $D$'s are molecular or Brownian diffusivities; $\mathbf{d}$'s describe any drift velocities relative to the fluid velocities that the trace constituent might possess (e.g., caused by gravity or diffusiophoresis); and $G$ and $L$ are gain and loss rates per unit volume caused by such processes as radioactive production and decay, chemical conversion, particle coagulation, and precipitation scavenging. To complete the specification of the problem, in addition to the governing Eqs. (4.1) and (4.2), coupling boundary conditions at the air–sea interface must be specified. One inviolable condition is equality of normal components of the fluxes in the two media. The required second interfacial boundary condition depends on characteristics of the trace constituent; examples will be given in a subsequent paragraph.

The continuity equations and coupling boundary conditions describe the entire problem of interest here, but this formalism exceeds present analysis capabilities. Consequently, the objective of efforts in this field of study is to develop approximations, or parameterizations, for terms in these equations and for the boundary conditions in forms suitable for application at space and time scales of interest. In turn, a major difficulty in this endeavor arises from the plethora of space and time scales contained in the equations; for example, Figure 4.1 qualitatively illustrates some of the vertical space scales and associated time scales contained in the equations. In this introductory section the essentials of this parameterization will be illustrated by averaging Eqs. (4.1) and (4.2) over progressively larger scales.

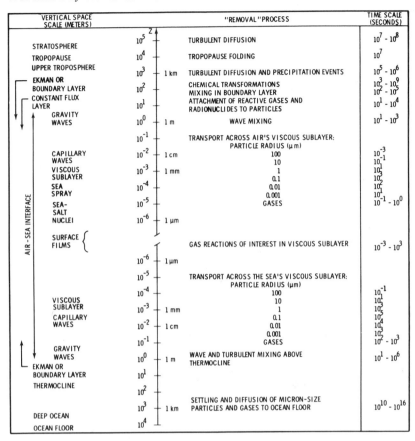

| VERTICAL SPACE SCALE (METERS) | | "REMOVAL" PROCESS | TIME SCALE (SECONDS) |
|---|---|---|---|
| STRATOSPHERE | $10^5$ | TURBULENT DIFFUSION | $10^7 - 10^8$ |
| TROPOPAUSE | $10^4$ | TROPOPAUSE FOLDING | $10^7$ |
| UPPER TROPOSPHERE | $10^3$ — 1 km | TURBULENT DIFFUSION AND PRECIPITATION EVENTS | $10^5 - 10^6$ |
| EKMAN OR BOUNDARY LAYER | $10^2$ | CHEMICAL TRANSFORMATIONS MIXING IN BOUNDARY LAYER | $10^3 - 10^9$ $10^0 - 10^0$ |
| CONSTANT FLUX LAYER | $10^1$ | ATTACHMENT OF REACTIVE GASES AND RADIONUCLIDES TO PARTICLES | $10^1 - 10^4$ |
| GRAVITY WAVES | $10^0$ — 1 m | WAVE MIXING | $10^1 - 10^3$ |
| | $10^{-1}$ | TRANSPORT ACROSS AIR'S VISCOUS SUBLAYER: PARTICLE RADIUS ($\mu$m) | |
| CAPILLARY WAVES | $10^{-2}$ — 1 cm | 100 | $10^{-3}$ $10^{-1}$ |
| VISCOUS SUBLAYER | $10^{-3}$ — 1 mm | 10 1 | $10^1$ $10^2$ |
| SEA SPRAY | $10^{-4}$ | 0.1 0.01 | $10^2$ $10^1$ |
| SEA-SALT NUCLEI | $10^{-5}$ | 0.001 GASES | $10^{-1} - 10^0$ |
| | $10^{-6}$ — 1 $\mu$m | | |
| SURFACE FILMS | | GAS REACTIONS OF INTEREST IN VISCOUS SUBLAYER | $10^{-3} - 10^3$ |
| | $10^{-6}$ — 1 $\mu$m | | |
| | $10^{-5}$ | TRANSPORT ACROSS THE SEA'S VISCOUS SUBLAYER: PARTICLE RADIUS ($\mu$m) | |
| | $10^{-4}$ | 100 | $10^{-1}$ |
| VISCOUS SUBLAYER | $10^{-3}$ — 1 mm | 10 1 | $10^1$ $10^5$ |
| CAPILLARY WAVES | $10^{-2}$ — 1 cm | 0.1 0.01 | $10^4$ $10^3$ |
| | $10^{-1}$ | 0.001 GASES | $10^2 - 10^3$ |
| GRAVITY WAVES | $10^0$ — 1 m | WAVE AND TURBULENT MIXING ABOVE THERMOCLINE | $10^1 - 10^6$ |
| EKMAN OR BOUNDARY LAYER | $10^1$ | | |
| THERMOCLINE | $10^2$ | | |
| | $10^3$ — 1 km | SETTLING AND DIFFUSION OF MICRON-SIZE PARTICLES AND GASES TO OCEAN FLOOR | $10^{10} - 10^{16}$ |
| DEEP OCEAN | | | |
| OCEAN FLOOR | $10^4$ | | |

FIGURE 4.1 An indication of the range of vertical space scales that are contained in Eqs. (4.1) and (4.2) and the estimates for the corresponding time scales for pollutant transfer.

A. MICROSCALE PARAMETERIZATION

As a first step, Eqs. (4.1) and (4.2) are averaged over what can be qualitatively described as microscale processes. These include not only the molecular-induced motions of the trace constituent but also the motions induced by high-frequency ($\gtrsim 1$ cycle per minute) turbulence. Let averages over these microscale processes be identified by tildes, e.g., $\tilde{\chi}$. Then Eqs. (4.1) and (4.2) become

$$\frac{\partial \tilde{\chi}}{\partial t} + \tilde{\mathbf{v}}_g \cdot \nabla \tilde{\chi} = -\nabla \cdot (\tilde{\mathbf{F}}_g + \tilde{\mathbf{d}}_g \tilde{\chi}) + \tilde{G}_g - \tilde{L}_g, \qquad (4.3)$$

$$\frac{\partial \tilde{C}}{\partial t} + \tilde{\mathbf{v}}_l \cdot \nabla \tilde{C} = -\nabla \cdot (\tilde{\mathbf{F}}_l + \tilde{\mathbf{d}}_l \tilde{C}) + \tilde{G}_l - \tilde{L}_l, \qquad (4.4)$$

where the microscale average flux caused by turbulent (as well as molecular) fluctuations is

$$-\tilde{\mathbf{F}}_g = -\widetilde{\mathbf{v}_g'\chi'} - \widetilde{\mathbf{d}_g'\chi'} + D_g \nabla \tilde{\chi}, \qquad (4.5)$$

in which the prime (') symbolizes the portion of the variable that fluctuates on the microscale but whose microscale average is zero; the correlation between drift and concentration fluctuations, $\widetilde{\mathbf{d}_g'\chi'}$, is retained for possible application to the case of sufficiently massive aerosol particles that do not exactly follow the fluid's turbulent fluctuations; and, again, $D$ is the molecular or Brownian diffusivity of the species. In analogy with this molecular diffusion term, a common parameterization of the turbulent flux is

$$-\widetilde{\mathbf{v}_g'\chi'} = \tilde{K} \cdot \nabla \tilde{\chi}, \qquad (4.6)$$

where $\tilde{K}$ is known as the (second-order tensor) turbulent or eddy diffusivity. Much has been written about the range of validity of this parameterization (e.g., see Corrsin, 1974); its use near the air–sea interface will be discussed in a later section.

Associated with these microscale-averaged continuity equations are parameterized boundary conditions. The condition for equality of normal components of fluxes at the interface ($z = z_i$) becomes

$$(\tilde{F}_{ng} + \tilde{d}_{ng}\tilde{\chi})\big|_{z=z_i} = (\tilde{F}_{nl} + \tilde{d}_{nl}\tilde{C})\big|_{z=z_i}. \qquad (4.7)$$

A major thrust of this review will be to discuss methods used to parameterize the required second boundary condition; for now, we display in Table 4.1 some of the many parameterizations available, each of which will be discussed later. In general, the major unsolved problems in this area of research are to interpret these parameterizations in terms of underlying microscale physical and chemical processes that dictate transfer across the interface.

B. ATMOSPHERIC RESIDENCE TIMES

To describe the flux of atmospheric trace constituents through the air–sea interface, it is usually an essential first step, as illustrated in the previous subsection, to average over, and to parameterize, the microscale processes. However, this is only the first step, and the importance and the required accuracy of the parameterizations can be seen only by illustrating how they can be applied and by noting what other approximations are introduced in the course of their application. In turn, this depends on the time and space scales of interest. For example, the microscale parameterizations could be used directly in detailed atmospheric models to determine the outcome of a specific tracer release or to assist in the design and execution of field experiments to measure wet or dry removal or to test their parameterizations. In contrast, for the predictions of long-term average pollutant fluxes to the oceans, it is expected that coarse models of large-scale meteorological processes will continue to be profitably applied, and it is therefore useful to see how the microscale parameterizations are used in such schemes.

One of the coarse models for large-scale meteorological processes utilizes the concept of atmospheric residence time. To introduce this concept qualitatively, consider the fate of a stable tracer released to the atmosphere and integrate Eq. (4.3) over the volume $V$ occupied by the tracer at time $t$. If the total amount of tracer still present in the atmosphere is $q$, then Eq. (4.3) can be used to define an atmospheric residence time, $\tau$, according to

$$\tau^{-1} = -\frac{1}{q}\frac{\partial q}{\partial t} = \frac{1}{q}\int_{V}[\nabla \cdot (\tilde{\mathbf{F}} + \tilde{\mathbf{d}}\tilde{\chi}) + L]\, dV, \qquad (4.8)$$

where the gain term has been ignored, $V$ is large enough so that there is no flux of the tracer through the volume's surface carried by the mean wind, and the subscript $g$ (for gas phase) on the flux and loss terms has been dropped for convenience. That the first equality in Eq. (4.8) defines a residence time can be seen qualitatively by considering the scenario illustrated in Figure 4.2. In Figure 4.2(a) between $t_0$ and $t_1$ it is assumed that only dry deposition depleted the tracer; between $t_1$ and $t_2$, wet deposition occurred; etc. Given these "data" for $q$ in Figure 4.2(a), the first equality in Eq. (4.8) can be used to evaluate the (time-dependent) residence time $\tau$ shown in Figure 4.2(b). A histogram that shows the distribution of the different residence times of Figure 4.2(b) is shown in Figure 4.2(c). In turn, it can be imagined that if this type of experiment were repeated a large enough number of times, then a

TABLE 4.1 Optional Ways to Specify the Second Interfacial Boundary Condition for Equations (4.3) and (4.4) for Dry Flux Past the Interface

| Boundary Conditions[a] | Remarks[b] |
|---|---|
| A. *Relate Concentrations* | |
| 1. For Gases | |
| (i) $\tilde{\chi}_i = H\tilde{C}_i$ | $H$ = Henry's law constant |
| (ii) $\tilde{\chi}_i = 0$ | Perfect sink (this condition decouples the equations) |
| 2. For Particles | |
| (i) $\tilde{\chi}_i = J\tilde{C}_i$ | $J$ = concentration jump[1] (related to the resuspension factor) |
| (ii) $\tilde{\chi}_i = 0$ | Perfect sink (see comment for similar case for gases) |

## 1. For Gases ($\tilde{d}_z = 0$)

(i)    $-\tilde{F}_z = k_g\,(\tilde{\chi}_b - \tilde{\chi}_i)$

(ii)    $-\tilde{F}_z = k_l\,(\tilde{C}_i - \tilde{C}_b)$

(iii)   $-\tilde{F}_z = k_t\,(\tilde{\chi}_b - H\tilde{C}_b)$

## 2. For Particles

(i)    $-(\tilde{F}_z + \tilde{d}_z\tilde{\chi})_{z_i} = v_d{}^{C}\tilde{\chi}_i$

(ii)    $-(\tilde{F}_z + \tilde{d}_z\tilde{\chi})_{z_i,h} = v_d{}^{Ch}\tilde{\chi}_b$

(iii)   $-(\tilde{F}_z + \tilde{d}_z\tilde{\chi})_{z_i} = v_d\tilde{\chi}_i - v_r\tilde{C}_i$

$k_g$   = gas-phase transfer velocity[2]

$k_l$   = liquid-phase transfer velocity

$k_t{}^{-1} = k_g{}^{-1} + Hk_l{}^{-1}$, total transfer velocity

$v_d{}^{C}$   = Calder's deposition velocity[3]

$v_d{}^{Ch}$ = Chamberlain's deposition velocity[4]

$v_r$   = resuspension velocity[1]

[a] Subscripts: $i$, interface; $b$, bulk concentration, measured at a convenient reference height, $h$.

[b] References:
[1] Slinn (1976a).
[2] Liss and Slater (1974).
[3] Calder (1961).
[4] Chamberlain (1960).

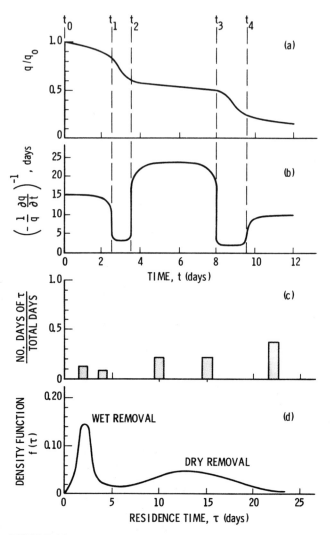

FIGURE 4.2   A schematic development (see text) of the probability density function $f(\tau)$ for the residence time, $\tau$.

density function for the residence time could be obtained, as shown in Figure 4.2(d). Finally, an (ensemble) average residence time for this type of tracer release could be defined as

$$\bar{\tau} = \int_0^\infty \tau f(\tau) d\tau;$$  (4.9)

however, this does not permit one to assess the effects of wet and dry removal processes separately on the residence time (and, possibly, the effects of other removal processes for other than nonreactive tracers).

What is desired now is to demonstrate the relationship between the atmospheric residence time and parameterizations for wet and dry removal processes in order that other needed parameterizations can be displayed. Toward this goal, we first average Eq. (4.8) over many tracer releases or over a suitably long time during the continuous release of a single tracer. Thus we ignore any distinction between ensemble and time averaging, although this point deserves further thought and discussion. The long-term average of large-scale meteorological processes will be denoted by an overbar, e.g., $\bar{\chi}$. Next, with use of the divergence theorem, the volume integral of the divergence of the dry flux is written as a surface integral of the flux. In the case of aerosol particle removal, the dry flux can be parameterized as in Table 4.1, e.g., as $v_d \bar{\chi}_b$, where $\bar{\chi}_b$ is the pollutant's average bulk air concentration at a convenient reference height (e.g., 1 or 10 m) and $v_d$ is a dry deposition velocity. Also for aerosol particles, the loss rate per unit volume by precipitation scavenging can be written as $\bar{L} = \bar{\psi \chi}$, where $\bar{\psi}$ is an average precipitation scavenging rate coefficient. With these parameterizations the average of Eq. (4.8) yields

$$\bar{\tau}^{-1} = \bar{\tau}_d^{-1} + \bar{\tau}_w^{-1},$$  (4.10)

where $\bar{\tau}_w^{-1} = \bar{\psi}$ and $\bar{\tau}_d^{-1} = v_d / \bar{h}_d$ in which $\bar{h}_d$ is a parameterization of the pollutant's long-term average vertical distribution

$$\bar{h}_d = \chi_b^{-1} \int_0^\infty \chi \, dz.$$  (4.11)

For example, if $\bar{h}_d = 3$ km and $v_d = 0.3$ cm sec$^{-1}$, then $\bar{\tau}_d \approx 10$ days; for a monthly average rainfall rate of 10 cm month$^{-1}$ and reasonable values for other terms in $\bar{\psi}$, it will be seen that for particles $\bar{\tau}_w$ is also of the order of 10 days.

Although this presentation is sketchy, it is hoped that the major point can be seen. It is that, toward the goal of predicting air-pollution fluxes to the oceans, obtaining microscale parameterizations of the removal processes is only the first step. Details of additional steps depend on the space and time scales of interest. For example, in the above discussion of atmospheric residence times, it is seen that while it is important to obtain accurate parameterizations of the dry deposition velocity, it is just as important to have accurate knowledge about average meteorological conditions such as those parameterized by $\bar{h}_d$. Similarly, it is necessary to know some average properties about precipitation. Although these needs are significant and will be discussed in somewhat greater detail in later sections, there is a tendency in this chapter to emphasize the microscale parameterizations; in contrast, Chapter 3 emphasized some of the needed parameterizations for large-scale meteorological processes.

### C.   GLOBAL-SCALE PARAMETERIZATIONS AND RESERVOIRS

Still coarser parameterizations are valuable, e.g., to estimate the global fate of pollutants released from sources in the 30–60° N latitude zone. Suitable parameterizations of the removal processes can be obtained by averaging Eq. (4.3) over reservoirs of desired size. Toward this end, let $Q_i$ be the total amount of the trace constituent of interest in reservoir $i$ and $F_i$ be the net outflow through the reservoir's surface. Further, if this net outflow is divided into an outflow, $O_i$, and an inflow, $I_i$, then clearly an average of Eq. (4.3) over reservoir $i$ leads to

$$\frac{dQ_i}{dt} = I_i - O_i + P_i - D_i, \qquad (4.12)$$

where $P_i$ and $D_i$ are the total production and destruction, respectively, of $Q_i$ in reservoir $i$. At steady-state conditions, Eq. (4.12) yields the transparent result

$$I_i + P_i = D_i + O_i, \qquad (4.13)$$

and at these conditions a similarly transparent reservoir residence time can be defined as

$$\tau_i = \frac{Q_i}{P_i + I_i} \equiv \frac{Q_i}{D_i + O_i}. \qquad (4.14)$$

Table 4.2 illustrates Eq. (4.14) for the case of precipitable water, although it should be emphasized that these results may be inaccurate because steady-state conditions rarely prevail and because both the mixing between latitude belts and the "dry (gaseous) deposition" of water vapor have been ignored.

Reservoir residence times can also be defined for unsteady conditions for the important special case for which the exchange and destruction processes are first order in $Q_i$. Then Eq. (4.12) becomes

$$\frac{dQ_i}{dt} = \sum_j{}' I_{ij}Q_j - \left(\sum_j{}' l_{ij}\right) Q_i + P_i - \lambda_i Q_i, \qquad (4.15)$$

where $I_{ij}$ and $l_{ij}$ are inflow and outflow (or *l*eaving) exchange rates; $\lambda_i$ are decay rates, and the prime on the summation sign symbolizes no sum on $i$. If $P_i$ is independent of $Q_i$, then the solution to Eq. (4.15) is that the steady-state concentration,

$$(Q_i)_{ss} = \frac{P_i + \sum_j{}' I_{ij}Q_j}{\lambda_i + \sum_j{}' l_{ji}}, \qquad (4.16)$$

is approached exponentially with an $e$-fold (residence, reservoir, or turnover) time constant

$$\tau_i = \left(\lambda_i + \sum_j{}' l_{ji}\right)^{-1}. \qquad (4.17)$$

It is noted that for steady-state conditions the two residence times, Eqs. (4.14) and (4.17), are equivalent.

A simple example using this formalism will illustrate where further research could improve this macroscale parameterization. Consider the box model of the troposphere shown in Figure 4.3. The values for the reciprocals of the exchange rates shown in Figure 4.3 were chosen only for illustrative purposes, although they were guided by the model of Lal and Rama (1966). That the exchange rates for transfer in opposite directions between some adjacent reservoirs differ by an order of magnitude reflects the approximate, order-of-magnitude differences in air masses of the reservoirs. (Note that net transfer between reservoirs terminates when the mixing ratios, not the $Q$'s, are equal.) If these exchange rates are used in Eqs. (4.16) and (4.17), then the results of the calculation for the case that $\lambda_i = 0$ and the only source, $P_i$, is in

**TABLE 4.2** Average Residence Time of Water Vapor in the Atmosphere as a Function of Latitude[a]

| | Latitude Range (degrees) | | | | | | | | |
|---|---|---|---|---|---|---|---|---|---|
| | 0–10 | 10–20 | 20–30 | 30–40 | 40–50 | 50–60 | 60–70 | 70–80 | 80–90 |
| Average precipitable water (g/cm$^2$) | 4.1 | 3.5 | 2.7 | 2.1 | 1.6 | 1.3 | 1.0 | 0.7[b] | 0.45[b] |
| Average precipitation (g/cm$^2$ year) | 186 | 114 | 82 | 89 | 91 | 77 | 42 | 19 | 11 |
| Residence time (days) | 8.1 | 11.2 | 12.0 | 8.7 | 6.4 | 6.2 | 8.7 | (13.4) | (15.0) |

[a]Reprinted with permission from C. E. Junge (1963). *Air Chemistry and Radioactivity*, Academic Press, New York, p. 10.
[b]Values extrapolated.

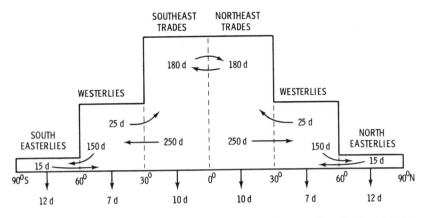

FIGURE 4.3 A schematic of the reservoirs chosen to illustrate Eqs. (4.16) and (4.17). The volumes of the reservoirs in each hemisphere are assumed to differ by an order of magnitude, and this explains why the reciprocals of some exchange rates also differ by an order of magnitude.

the 30–60° N reservoir, are as shown in Table 4.3 and Figure 4.4. The fallouts from the reservoirs were calculated from

$$(\text{Fallout})_i = l_{is} Q_i, \qquad (4.18)$$

in which, for Table 4.3, $l_{is}$ is the removal rate from the $i$th reservoir found by averaging over appropriate latitude zones of Table 4.2. In Figure 4.4 two cases are shown: for one, the $l_{is}$ of Table 4.2 were used; for the other, the removal rates were assumed to be a factor of 10 slower than for water vapor. This example illustrates the importance not only of accurate parameterizations of the removal processes but also of accurate parameterization of the exchange rates between reservoirs. Some appropriate research topics will be suggested for consideration in a later section.

### D.  SUMMARY AND COMMENTS

In summary, the objective of these introductory remarks was to display the unifying features of the continuity equations (4.1) and (4.2), the types of parameterizations needed, and the interrelations among progressively coarser descriptions of the behavior of atmospheric trace constituents. It is re-emphasized that parameterizations are needed not only for the removal mechanisms themselves but, depending on the

TABLE 4.3   Model Results for Fallout Rates as Given in Table 4.2 and Exchange Rates as Shown in Figure 4.3

| | Latitude | | | | | |
|---|---|---|---|---|---|---|
| Reservoir | 90–60° S | 60–30° S | 30–0° S | 0–30° N | 30–60° N | 60–90° N |
| $l_{is}$ (days) | 12 | 7 | 10 | 10 | 7 | 12 |
| $\tau_i$ (days) | 6.7 | 5.2 | 9.1 | 9.1 | 5.2 | 6.7 |
| $P_i$ (units/day) | 0 | 0 | 0 | 0 | 1 | 0 |
| $Q_i$ (units) | $9.4 \times 10^{-5}$ | $2.1 \times 10^{-3}$ | $1.0 \times 10^{-1}$ | $2.0 \times 10^{0}$ | $5.4 \times 10^{0}$ | $2.4 \times 10^{-1}$ |
| $[\text{Fallout}]_i$ (units/day) | $7.9 \times 10^{-6}$ | $3.0 \times 10^{-4}$ | $1.0 \times 10^{-2}$ | $2.0 \times 10^{-1}$ | $7.7 \times 10^{-1}$ | $2.0 \times 10^{-2}$ |

space and time scales of interest, also for (1) the correlations between velocity and concentration fluctuations, $\widetilde{v'\chi'}$; (2) the turbulent (or eddy) diffusivity, $\tilde{K}$; (3) the effective height $\bar{h}_d$ and average removal rate $\bar{\psi}$; or (4) interreservoir exchange rates. In this chapter the emphasis will be on parameterizing microscale aspects of the wet and dry removal processes.

Incidentally, we cannot side with the critics who view reservoir models as outmoded and oversimplified; we are certain that these models and the concept of reservoir residence times will continue to play an essential role in developing useful predictions of material fluxes to the oceans. It is noted from Eq. (4.10), (4.14), and (4.17) that residence times are essentially the inverses of the removal rates, and partitioning the atmosphere (and other geospheres) into various reservoirs essentially amounts to obtaining the first few terms in a spatial Fourier analysis of the removal rates. It is true that, in the limit of infinitely many reservoirs, this partitioning will return the analysis to the continuum formulation from which it originated, but it is also apparent that rarely would this much detail be desired. An exception, perhaps, would be the case for which it is desired to describe the fate of a specific pollutant or tracer release. In contrast, residence-time concepts can provide simple, convenient, and first-order estimates for time-averaged fluxes to the ocean.

With this said, however, we now turn to our primary task of discussing microscale parameterizations for wet and dry removal processes. The importance of this task is illustrated in Figure 4.4, where it is seen that, if the removal rates for a pollutant injected into the 30–60° N reservoir are reduced by an order of magnitude, then the

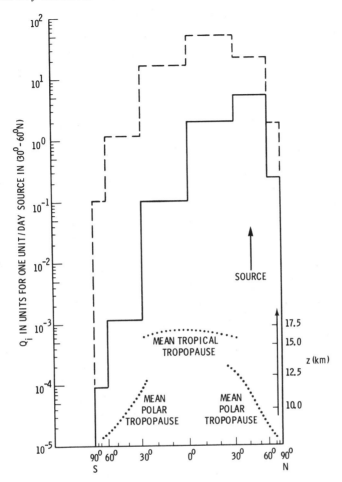

FIGURE 4.4   The solutions to the set of Eqs. (4.16) and (4.17) for the steady-state concentration of pollutant in each reservoir for the case of a unit source in the reservoir 30–60° N. The solid curve is for the removal and transfer rates shown in Figure 4.3; the dashed curve is for the fallout rates one tenth of the rates shown in Figure 4.3; the dotted curve is an indication of the mean tropopause [for January–August 1959, see Junge (1963), p. 246]. With the latitude and altitude scales shown, the same area under the tropopause curve represents the same mass fraction of the atmosphere, and, therefore, this plot supports the approximation made in formulating the problem that the air masses in the three reservoirs differ by about an order of magnitude.

pollutant's concentration in, for example, the south polar zone is increased by about *three* orders of magnitude. In the next section, the emphasis is on precipitation scavenging; following that will be a description of parameterizations for dry deposition; finally there is a survey section on current and suggested future research.

## III. PRECIPITATION SCAVENGING

The term precipitation scavenging is used here to describe atmospheric trace constituent removal from the atmosphere by various types of precipitation. The term precipitation refers to those hydrometeors (raindrops, snowflakes, etc.) whose gravitation terminal velocity, $v_t$, exceeds local updraft speeds by about 10 cm sec$^{-1}$. Thus, the incorporation of air pollutants into cloud droplets ($v_t \simeq 1$ cm sec$^{-1}$) will not be described as precipitation scavenging, since this stage does not ensure the pollutant's removal from the atmosphere. Indeed, even the capture and subsequent transport of materials by precipitation may not lead to their removal if the hydrometeors evaporate en route to the earth's surface; in this case, the process is just one of many ways that substances can be redistributed in the atmosphere. In this section, some theoretical considerations will be introduced and then used to interpret some especially useful data; in a later section, some precipitation scavenging research topics will be suggested.

### A.   WET REMOVAL TIME SCALE

In the previous section, an average wet removal time scale, a contribution to the residence time for particles, was given in Eq. (4.10) as $\bar{\tau}_w = \bar{\psi}^{-1}$, where $\bar{\psi}$ is an average precipitation-scavenging rate coefficient. The main thrust of this section will be to present parameterizations for the removal rate, $\psi$; but before doing this it may be informative to see an alternative derivation of the wet removal time scale. This can be seen by envisaging the fate of a specific, stable tracer released to the atmosphere. If dry deposition is ignored, then the total amount of tracer remaining in the atmosphere at any time $t$ might be as shown in Figure 4.5, curve (a). Thus, after a number of days (say, $\bar{\nu}^{-1}$ days on the average) the tracer is presumed to interact with a precipitating storm system and a fraction $\epsilon_1$ of the tracer (say, $\bar{\epsilon}$ on the average) is removed by the storm, leaving the fraction $(1 - \epsilon_1)$ still airborne. Continuing in this manner we see that, after $n$ storms, the

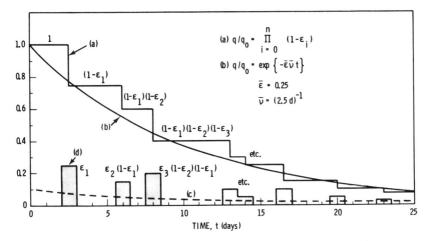

FIGURE 4.5 A schematic development (see text) of a continuum description for precipitation scavenging.

fraction of the tracer still airborne is

$$\prod_{i=0}^{n} (1 - \epsilon_i).$$

The corresponding deposition pattern is shown as curve (d) in Figure 4.5. It can be envisaged that if this type of experiment were repeated a large enough number of times, then the average curves (b) and (c) in Figure 4.5 would be obtained.

Analytically, the average concentration can be found from the approximation

$$\frac{\overline{q}(t)}{q(o)} = \prod_{i=0}^{n} \overline{(1 - \epsilon_i)} \simeq \left(1 - \frac{\overline{\epsilon}\overline{v}t}{n}\right)^n \simeq \exp(-\overline{\epsilon}\overline{v}t) \quad (4.19)$$

or from assuming that precipitation scavenging is a Poisson process and using known properties of Poisson processes or from integrating the analytical statement that during $\Delta t$ the average amount removed is $-\Delta q = (\overline{v}\Delta t)\overline{\epsilon}q$. Similarly, the average amount of tracer deposited during $\Delta t$ is $\overline{\epsilon}vq\Delta t$, and therefore the rate of wet deposition is

$$\overline{\epsilon}\overline{v}q = \overline{\epsilon}\overline{v}q_o \exp(-\overline{\epsilon}\overline{v}t). \quad (4.20)$$

Equations (4.19) and (4.20) describe the average curves (b) and (c) of Figure 4.5 and display a second formulation for the characteristic wet removal time scale

$$\bar{\tau}_w = \bar{\psi}^{-1} = (\bar{\epsilon}\bar{\nu})^{-1}. \tag{4.21}$$

Although this simple analysis is informative, there are a number of impediments to using it for quantitative descriptions of precipitation scavenging. For example, it has been implicitly assumed that the fraction of the material removed by an individual storm, $\epsilon$, is independent of the amount of material present. For most material this would be true, but cases for which it might fail include the following: if there are too many particles present that act as cloud-droplet or ice-crystal nuclei then their presence could influence the precipitation rate; for some gases, cloud droplets can become saturated with the gas and then only a certain amount, not a specific fraction, of the gas will be removed.

Of more significance, however, is the present lack of knowledge about $\bar{\epsilon}$ and $\bar{\nu}$. To separate the problem into its microscale and larger-scale portions, consider the case of aerosol particles, which, for reasons described later, are typically more efficiently removed from within rather than from below clouds. Then let the average fraction of the aerosol particles removed by a storm be written as $\bar{\epsilon} = \bar{\epsilon}_i\bar{\epsilon}_{cw}$, where $\bar{\epsilon}_i$ is the average fraction of the particles incorporated into the cloud water and $\bar{\epsilon}_{cw}$ is the average efficiency with which cloud water is removed by a storm. Then the meteorological problems are to obtain $\bar{\epsilon}_{cw}$ and the average time, $\bar{\nu}^{-1}$, between encounters with precipitation storms. These parameters are not known at all well, even over land. For example, clearly $\epsilon_{cw}$ can vary from zero (for a nonprecipitating storm) to perhaps 90 percent; for precipitating storms, probably $\bar{\epsilon}_{cw} \simeq 1/3$ is accurate to within a factor of 3, but there are few data to support this contention. Similarly, $\bar{\nu}$ can be estimated to be the total precipitation at a site (say, 100 cm yr$^{-1}$) divided by the average precipitation per storm (say, 1 cm per storm) or $\bar{\nu} \simeq (3\ \mathrm{days})^{-1}$, but clearly this is a crude approximation. Even the annual average precipitation to the oceans is not known well. With these crude estimates follows the equally crude estimate of wet removal's contribution to the average residence time for water vapor: $\bar{\tau}_w \simeq (\bar{\epsilon}_{cw}\bar{\nu})^{-1} \simeq 10$ days. This discussion illustrates the need for additional meteorological research to obtain $\bar{\epsilon}_{cw}$ and $\bar{\nu}$; the discussion in the next subsection will illustrate required research for the determination, essentially, of $\bar{\epsilon}_i$.

## B. WET REMOVAL OF PARTICLES

Although the above analysis is informative, fortunately there is another way to estimate $\bar{\tau}_w$ quantitatively, i.e., $\bar{\tau}_w = \overline{\psi}^{-1}$. For aerosol particles of radii $a$ irreversibly captured by precipitation, the loss rate per unit volume in Eq. (4.3) varies linearly with the air concentration of the particles: $\tilde{L}(a) = \tilde{\psi}\tilde{\chi}(a)$, where $\tilde{\psi}$ is the microscale-average removal rate. This removal rate can be obtained from a simple analysis of collisions between particles and hydrometeors (Chamberlain, 1960; Englemann, 1968). The familiar result (which, it should be noted, is valid both within and below clouds) is

$$\tilde{\psi}(\mathbf{r},t;a) = \int_0^\infty dl N(\mathbf{r},t;l)v_t(l)AE_1(a,l), \qquad (4.22)$$

where $N(l)dl$ is the number of hydrometeors per unit volume whose length scale (e.g., drop radius) is between $l$ to $l + dl$; $v_t (\gtrsim 10 \text{ cm sec}^{-1})$ is the hydrometeors' gravitational terminal velocity and $A$ is their cross-sectional area; and $E_1(a,l)$ is the collection efficiency (or $AE_1$ is the effective collection cross section). In turn, the collection efficiency can be written as a product of the collision efficiency and a retention efficiency, usually taken to be unity. Figures 4.6 and 4.7 show semi-empirical expressions for the collision efficiencies between particles and raindrops (Figure 4.6) or snowflakes (Figure 4.7), which correlate most of the available data (Slinn, 1976b).

Because of remaining uncertainties in the collection efficiencies and because of lack of *a priori* data on hydrometeor size distributions, Slinn (1976c) advanced the following approximations to Eq. (4.22):

$$\psi_r(\mathbf{r},t;a) = c \, \frac{p(\mathbf{r},t)}{R_m} \, E_1(a,R_m), \text{ for rain;} \qquad (4.23)$$

$$\psi_s(\mathbf{r},t;a) = g \, \frac{\rho_w}{\rho_a} \, \frac{p(\mathbf{r},t)}{v_{tm}^2} \, E_2(a,\lambda), \text{ for snow.} \qquad (4.24)$$

These approximations were obtained, essentially, by multiplying and dividing the integrand in Eq. (4.22) by $l$ and recalling the definition of the precipitation rate. In these equations, $c$ is a numerical factor $\simeq$ 1/2; $p(\mathbf{r},t)$ is the precipitation rate (rainwater equivalent); $R_m$ is the volume–mean drop radius; $g$ is the acceleration of gravity; $\rho_w$ and $\rho_a$ are the mass densities of water and air, respectively; $v_{tm}$ is an average terminal velocity for the snowflakes, and $E_1$ and $E_2$ are average

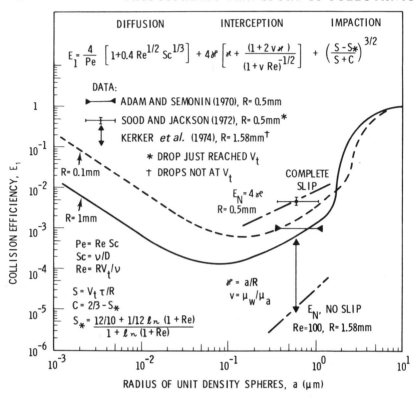

FIGURE 4.6 Semiempirical expressions for the collision efficiency between drops and particles as a function of particle size and accounting for diffusion, interception, and inertial impaction. The diffusion and impaction portions of the curves have sufficient experimental support to consider them reliable to within a factor of 2 or 3. See Slinn (1976c) for a possible explanation of the scatter in the data for particles of radii ~0.5 $\mu$m; see Slinn (1976b) for an illustration of the importance of accounting for aerosol particle growth by attachment to plume or cloud particles or by water-vapor condensation. Reprinted with permission from the *Journal of Air, Water and Soil Pollution.*

collection efficiencies, evaluated using the mean drop size and characteristic length $\lambda$ (see Figure 4.7), respectively.

From the expressions, Eqs. (4.23) and (4.24), for the removal rates it is easy to obtain expressions for the wet flux of aerosol particles to the earth's surface. If the particle concentration does not vary significantly over short horizontal distances, then an integral over the average hydrometeor's flight path can be replaced by a vertical integral. As a result, the wet flux is simply

$$W = \int_0^\infty \psi \tilde{\chi} dz \,, \qquad (4.25)$$

with $\psi$ as given in Eqs. (4.23) or (4.24). For example, for rain scavenging, Eq. (4.23) in (4.25) yields

$$W(\mathbf{r},t;a) = c\left[p_0(\mathbf{r},t)/R_{m0}\right] E_{1,0}(a,R_m) h_w \tilde{\chi}_b \,, \qquad (4.26)$$

where the subscript zero refers to surface level conditions; $\tilde{\chi}_b$ (or $\tilde{\chi}_0$) is the particle air concentration at a convenient reference height; and the

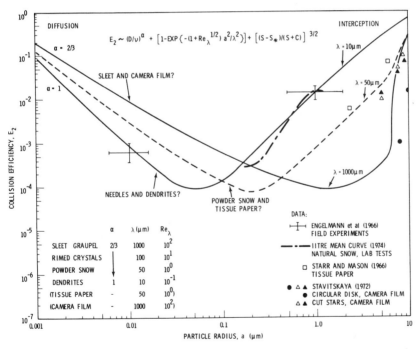

FIGURE 4.7 A tentative suggestion for the collision efficiency between particles and ice crystals. For discussions of theoretical analyses and experimental bases, see Slinn (1976c). Reprinted with permission from the *Journal of Air, Water and Soil Pollution*.

characteristic height $h_w$ from which the particles are removed has been defined by

$$h_w = \int_0^\infty \frac{p(z)}{p_0} \frac{\tilde{\chi}(z)}{\tilde{\chi}_b} \frac{E_1(a,R_m)}{E_{1,0}(a_0,R_{m0})} \, dz. \qquad (4.27)$$

Similar expressions can be obtained for snow scavenging using Eq. (4.24).

The collection efficiency is strongly dependent on particle size (see Figures 4.6 and 4.7), and since particle growth is usually substantial within clouds either because the particle acts as a cloud-droplet or ice-crystal nucleus or because of particle attachment to existing cloud particles, the range of integration in Eq. (4.27) can usually be restricted to the height of the active, rain-forming region within the cloud. For example, to fit data recently obtained by Gatz (1976a) and to be described below, Slinn (1976b) used $c = 1/2$, $R_{m0} = 0.5$ mm, $E_1$ as given in Figure 4.6 with a reasonable particle growth rate, and $h_w \simeq 600$ m.

In the case for which in-cloud scavenging is more important than below-cloud scavenging, the wet flux of particles from a steady-state cloud can be derived using the simple method developed by Junge (1963). In this case, if the average fraction $\langle \epsilon_i \rangle$ of the particles entering the cloud is incorporated into the cloud water, then the concentration of the captured particles is $\langle \epsilon_i \chi \rangle$ units per volume or $\langle \epsilon_i \chi \rangle / \langle L \rangle$ units per mass of cloud water, where $\langle \rangle$ represents an average value over the height of the storm and $\langle L \rangle$ is the average condensed water content of the clouds. If this concentration $\langle \epsilon_i \chi \rangle / \langle L \rangle$ also appears in the precipitation from the storm (which would be true for precipitation formed by gravitational coagulation of equally concentrated drops and where no evaporation or dilution occurred en route to the earth's surface), then the flux of material from the storm would be

$$W = \frac{\langle \epsilon_i \chi \rangle}{\langle L \rangle} \rho_w p_0, \qquad (4.28)$$

where $\rho_w p_0$ is the mass flux of water from the storm.

## C.  PARTICLE WASHOUT RATIOS

For comparisons with experimental data, it is useful to define the washout ratio, $r$, which is the ratio of the material's concentration in surface-level precipitation, $\kappa_o$, to its average concentration in surface level air, $\tilde{\chi}_o$ or $\tilde{\chi}_b$, where the bulk air concentration is measured at a convenient reference height. Since the flux of precipitation to the surface is $p_o$, then $\kappa_o = W/p_o$ and, using Eq. (26) or (28), the washout ratio becomes

$$r = \left(\frac{\kappa}{\chi}\right)_o = \frac{W}{p_o \chi_b} = \frac{ch_w E_{1,0}}{R_{mo}} = \left\langle \epsilon_i \frac{\rho_w}{L} \frac{\chi}{\chi_b} \right\rangle. \qquad (4.29)$$

For example, for $c = 0.5$, $R_m = 0.5$ mm, and $h_w = 500$ m or for $\rho_w = 1$ g cm$^{-3}$, $\langle L \rangle = 10^{-6}$ g m$^{-3}$, and $\langle \chi/\chi_b \rangle = 0.5$, then Eq. (29) becomes

$$r = 0.5 \times 10^6 E_{1,0} = 0.5 \times 10^6 \langle \epsilon_i \rangle. \qquad (4.30)$$

Usually the washout ratios are reported for a particular element or compound, independent of particle size, in which case Eq. (29) should be averaged over the particle size distribution of the species.

Table 4.4 shows some measured washout ratios. They are seen to be generally in the range $10^5$–$10^6$. Incidentally, some authors report dimensional washout ratios (with dimensions such as m$^3$ cm$^{-3}$, kg of air/liter of water, or m$^3$ liter$^{-1}$), and, therefore, the reader may find washout ratios in the literature near $10^0$ or $10^3$ rather than $10^6$. The dependence of the washout ratios on mass-median size of the particles, shown in Table 4.4 and reported by Gatz (1976a), can be fit with the theory leading to Eq. (26) (Slinn, 1976b). Actually, the value of the washout ratio near $10^6$ (which is a physically real concentration factor) can be seen from a simple physical argument that all the particles in an air volume of about $10^4$ cm$^3$ are transferred to, say, $10^6$ 10-$\mu$m cloud droplets and that these droplets coagulate to form one 1-mm drop; then the particles initially distributed in a volume of $10^4$ cm$^3$ will be concentrated into a volume of about $10^{-2}$ cm$^3$. It is this selective condensation plus coagulation that accounts for the frequently observed washout ratio (or precipitation scavenging concentration factor) of $10^6$. Using these measured washout ratios, the wet flux of aerosol particles to the ocean can be estimated as

$$W = r p_o \tilde{\chi}_b. \qquad (4.31)$$

However, it should be noted that none of the data in Table 4.4 have been obtained at an ocean site. Clearly it would be useful to obtain such data, and, to assist in its interpretation, it would be valuable if simultaneous measurements were obtained of particle size, the pollutant's

TABLE 4.4  Some Measured Washout Ratios, $r = (\kappa/\chi)_0$

| Scavenged Material | | $r \times 10^{-6}$ | Notes |
|---|---|---|---|
| Fallout radionuclides | | 1.0 ± 0.3 | 1957–1965 average for many radionuclides sampled at locations throughout the world; inorganic dust data from continuous measurements (Gedeonov et al., 1970) |
| Mass of inorganic dust | | 3.2 | |
| Total pollen (~15–50-μm diam) | | 0.65–3.8 | Scavenged by convective storms in Oklahoma (Gatz, 1966 and personal communications, 1976) |
| Aerosol particles | | | |
| Element | mmd (μm) | | |
| Pb | ~0.5 | 0.063 | Scavenged by convective storms near St. Louis; reported data multiplied by $10^6$ cm$^3$ m$^{-3}$/1.2 $\times$ 10$^3$ g of air m$^{-3}$ to convert to dimensionless form; mean washout ratios for all 1971–1972 data (Gatz, 1976a) |
| Zn | ~1.0 | 0.15 | |
| Fe | ~3.0 | 0.21 | |
| Ca | — | 0.29 | |
| Mn | ~2.0 | 0.31 | |
| Mg | 5–7 | 0.38 | |
| K | — | 0.46 | |
| | Avg. $v_d$ (cm sec$^{-1}$) | | |
| Cl | 0.06 | 1.08 | Annual average values, July 1972–June 1973, at Harwell, U.K.; reported data divided by 1.2 $\times$ 10$^{-3}$ g of air cm$^{-3}$ to convert to dimensionless form; average dry deposition velocities deduced from dry flux to a mechanical collector; rain sampler was automatically covered during dry periods; some local contamination may explain the high value of $v_d$ for Ni. Data obtained |
| Zn | 0.07 | 0.40 | |
| Br | 0.08 | 0.17 | |
| Na | 0.1 | 0.98 | |
| Se | 0.15 | 0.25 | |

| | | |
|---|---|---|
| Sb | 0.15 | 0.16 |
| V | 0.23 | 0.16 |
| Pb | 0.25 | 0.24 |
| Cs | 0.31 | 0.24 |
| Mn | 0.33 | 0.24 |
| Cr | 0.36 | 0.36 |
| As | 0.37 | 0.22 |
| Co | 0.50 | 0.85 |
| Fe | 0.94 | 0.23 |
| Al | 1.1 | 0.26 |
| Sc | 1.3 | 0.17 |
| ( Ni | 1.6 | 0.6 ) |

via personal communications from P.A. Cawse, Environmental and Medical Sciences Division, B.364, Atomic Energy Research Establishment, Harwell, Oxon OX11 ORA, U.K. Also, see Peirson et al. (1974)

Reactive gases

| | | |
|---|---|---|
| $[NH_4^+]\,/\,[NH_3]$ | 0.12–0.22 | Ratio taken of reaction product mass concentration in rain to precursor-gas mass concentration in air at various locations in Europe (Junge, 1963) |
| $[NO_3^-]\,/\,[NO_2]$ | 0.18–0.40 | |
| $[SO_4^{2-}]\,/\,[SO_2]$ | 0.05–0.15 | |
| $[SO_4^{2-}]\,/\,[SO_2]$ | 0.01–0.1 | Estimated from available data for St. Louis Metromex Project; convective storm scavenging |
| $[SO_4^{2-}]\,/\,[SO_2]$ | 0.15, 0.30 | Estimated from July 1972 to June 1973, National Air Sampling data for urban and nonurban sites, respectively, in Northeastern United States (Dana, 1976) |

vertical distribution, and properties of the precipitation (raindrops size and concentration, rain rates, etc.) and the storm (type, vertical and areal extent, duration, etc.).

### D.   GAS SCAVENGING

For gases that form simple solutions in water, the term in the continuity equation describing loss rate per unit volume must account for the possible desorption of the gas from raindrops:

$$\tilde{L} = \Lambda(\tilde{\chi} - H\tilde{\kappa}), \tag{4.32}$$

where $\Lambda$ is an average rate of gas transfer to the drops, $\tilde{\kappa}$ is the concentration of the gas in the drop, and $H$ is Henry's law constant. However, the transfer rate by molecular diffusion to drops is relatively rapid, and the small drop size or drop internal circulation promotes the rapid equilibration of the gas concentration in the drop (Postma, 1970; Hales, 1973); therefore, after a drop has fallen only a few tens of meters, typically it will become saturated with the gas. Consequently, except very near the source of a gas (e.g., within a few stack heights downwind of the source) it is usually an adequate approximation to set the surface-level concentration of an unreactive gas in raindrops equal to the equilibrium value:

$$\kappa_0 = H^{-1} \chi_0 = \alpha\chi_0, \tag{4.33}$$

where $\alpha \equiv H^{-1}$ is the solubility coefficient. This gives for the washout ratio and wet flux

$$r = (\kappa/\chi)_0 = \alpha; \qquad W = rp_0\chi_0. \tag{4.34}$$

In Table 4.5 are shown the Henry's law constants for a number of gases. It is seen that the resulting washout ratios for low-molecular-weight gases are substantially smaller than the values of $10^5$–$10^6$ typical for aerosol particles scavenged by either rain or snow. Snow scavenging of gases can generally be ignored unless the ice-crystal lattice is particularly accommodating for the gas molecules.

Table 4.5 includes some gases such as $CO_2$, $SO_2$, and $NH_3$ that are reactive in water; the corresponding washout ratios are for the wet removal only of the dissolved gases, not of their reaction products. To account for the details of the scavenging of reactive gases, it is necessary to evaluate the possible chemical reactions that these gases can undertake in the presence of other dissolved gases and a host of other trace constituents, some of which could catalyze specific reactions (see Chapter 8 for a more detailed discussion). These complicated analyses can lead to the definition of an enhanced solubility coefficient, $\alpha^*$ (e.g., see Junge, 1963; Postma, 1970; Hales, 1973); the enhanced solubility coefficient for $SO_2$ for two pH values and as a function of $\chi$ is shown in the last section of Table 4.5 (Dana, 1976). However, for the case of wet removal to the ocean, it would seem that a major simplification is available. Whereas clouds are effective "reaction vessels" for the chemical conversion of reactive gases to relatively involatile products (e.g., $SO_2 \rightarrow SO_4^{2-}$) and whereas most material would be expected to be subjected to at least ten condensation–evaporation cycles (i.e., clouds) before being deposited on the oceans (Junge, 1964), it would seem to be acceptable to ignore the wet removal of most reactive gases and, instead, to assume that reactive pollutant gases are converted to their reaction products before they are scavenged. The appropriateness of this assumption is strengthened by the large values of the washout ratios for the reactive gases listed in the last section of Table 4.4; as further support for this assumption, it is noted that these reactions products (i.e., $NH_4^+$, $NO_3^-$, and $SO_4^{2-}$) were measured in polluted atmospheres, possibly after experiencing only a single storm.

### E. SUMMARY AND COMMENTS

In this section, three different methods have been described that can supply estimates for the wet-removal time scale. From Eq. (4.21) or from Eq. (4.10) with Eq. (4.23) for rain [or Eq. (4.24) for snow] or with the washout ratio as given by Eq. (4.29) we have

$$\bar{\tau}_w \simeq \overline{(\epsilon_i \epsilon_{cw} \nu)} \simeq \bar{R}_m/(c\bar{p}_o \bar{E}_1) \simeq \bar{h}_w/(\overline{rp}_o). \tag{4.35}$$

For example, if for particles we take $\bar{\epsilon}_i \simeq 1$, $\bar{\epsilon}_{cw} = 0.25$, and $\bar{\nu} = (2.5$ day$)^{-1}$; or $\bar{R}_m = 0.25$ mm, $c = 0.5$, $\bar{p}_o = 100$ cm yr$^{-1}$, and $\bar{E}_1 = 10^{-2}$; or $\bar{h}_w = 10$ km and $\bar{r} = 0.3 \times 10^6$, then Eq. (4.35) yields $\bar{\tau}_w \simeq 10$ days. With these three different approaches available one would hope that the final estimates for the wet-removal time scale would be fairly reliable. Unfortunately, however, all three approaches contain significant uncertainties; none of the estimates is reliable to within a factor of 2 or 3, and they may be incorrect by as much as an order of magnitude. At present, the best available procedure to estimate the wet flux of pollutants to the ocean is probably to use the measured washout ratios

TABLE 4.5    Henry's Law Constants, $H$, Solubility Coefficients, $\alpha$, and Washout Ratios, $r^a$

| Compound | | $H$ or $\alpha^{-1}$ | $r = (\kappa/\chi)_o = \alpha$ |
|---|---|---|---|
| I. | CO | $3.73 \times 10^1$ | $2.7 \times 10^{-2}$ |
| | $H_2S$ | $3.22 \times 10^1$ | $3.1 \times 10^{-2}$ |
| | $CH_4$ | $2.57 \times 10^1$ | $3.9 \times 10^{-2}$ |
| | NO | $1.84 \times 10^1$ | $5.4 \times 10^{-2}$ |
| | $C_2H_6$ | $1.72 \times 10^1$ | $5.8 \times 10^{-2}$ |
| | $C_2H_4$ | $6.82 \times 10^0$ | $1.5 \times 10^{-1}$ |
| | $O_3$ | $2.19 \times 10^0$ | $4.6 \times 10^{-1}$ |
| | $N_2O$ | $1.26 \times 10^0$ | $8.0 \times 10^{-1}$ |
| | $CO_2$ | $9.30 \times 10^{-1}$ | $1.1 \times 10^0$ |
| | $C_2H_2$ | $8.23 \times 10^{-1}$ | $1.2 \times 10^0$ |
| | $SO_2$ | $7.85 \times 10^{-3}$ | $1.3 \times 10^2$ |
| | $NH_3$ | $4.71 \times 10^{-4}$ | $2.1 \times 10^3$ |
| II. | CO | $5.0 \times 10^1$ | $2.0 \times 10^{-2}$ |
| | $CH_4$ | $4.2 \times 10^1$ | $2.4 \times 10^{-2}$ |
| | $CCl_3F$ | $5 \ \ \times 10^0$ | $2.0 \times 10^{-1}$ |
| | $N_2O$ | $1.6 \times 10^0$ | $6.3 \times 10^{-1}$ |
| | $CCl_4$ | $1.1 \times 10^0$ | $9.1 \times 10^{-1}$ |
| | $(CH_3)_2S$ | $3.0 \times 10^{-1}$ | $3.3 \times 10^0$ |
| | $CH_3I$ | $2.4 \times 10^{-1}$ | $4.2 \times 10^0$ |
| | $SO_2$ | $3.8 \times 10^{-2}$ | $2.6 \times 10^1$ |

shown in Table 4.4 and, if necessary, to extrapolate from these data using the theory presented in this section to guide extrapolations. In a later section, some recommended research to improve our knowledge of wet removal processes will be described. Before doing this, however, it is useful to look at other important removal processes.

## IV.  DRY DEPOSITION

In Section II, the residence times of atmospheric trace constituents were related to the wet and dry removal processes. In Section III, the magnitude of the wet removal flux was written as $W = rp_o\tilde{\chi}_b$ and expressions for the washout ratio were obtained. In this section, the focus will be on the dry flux $D_d$, whose magnitude to a horizontal surface can be written as

$$D_d = -\left(\tilde{F}_z + \tilde{d}_z\tilde{\chi}\right), \tag{4.36}$$

TABLE 4.5
Continued

| Compound | $H$ or $\alpha^{-1}$ | $r = (\kappa/\chi)_o = \alpha$ |
|---|---|---|
| III.  Hg | $4.8 \times 10^{-1}$ | $2.1 \times 10^0$ |
| Polychlorinated biphenyls | | |
| Aroclor 1260 | $2.9 \times 10^{-1}$ | $3.4 \times 10^0$ |
| Aroclor 1248 | $1.5 \times 10^{-1}$ | $6.8 \times 10^0$ |
| Aroclor 1254 | $1.1 \times 10^{-1}$ | $8.8 \times 10^0$ |
| Aroclor 1242 | $2.4 \times 10^{-2}$ | $4.2 \times 10^1$ |
| Pesticides | | |
| DDT | $1.6 \times 10^{-3}$ | $6.2 \times 10^2$ |
| Aldrin | $5.3 \times 10^{-4}$ | $1.9 \times 10^3$ |
| Lindane | $2.0 \times 10^{-5}$ | $5.0 \times 10^4$ |
| Dieldrin | $7.7 \times 10^{-6}$ | $1.3 \times 10^5$ |
| IV.  Sulfur | $\chi$(ppbv) | $r = \alpha^*$ |
| pH = 5.0 | 1 | $3.7 \times 10^4$ |
| | 3 | $3.0 \times 10^4$ |
| | 10 | $2.2 \times 10^4$ |
| | 30 | $1.5 \times 10^4$ |
| pH = 6.0 | 1 | $9.0 \times 10^4$ |
| | 3 | $5.5 \times 10^4$ |
| | 10 | $3.2 \times 10^4$ |
| | 30 | $1.9 \times 10^4$ |

[a]The solubility coefficients typically decrease with increasing temperature (e.g., for $O_2$ the solubility in water at 20°C is about 60 percent of its value at 0°C) and typically decrease with the presence of other constituents (e.g., for $O_2$ the solubility in seawater is about 80 percent of its value in freshwater at the same temperature). The first group of data listed here, I, for freshwater at 15°C, is taken from Heines and Peters (1974); the second group, II, for seawater at an unspecified temperature is from Liss and Slater (1974); the third group, III, are estimates for pure water recently reported by Junge (1975); the fourth group, IV, is the enhanced solubility coefficient, $\alpha^*$, in moles per unit volume of water to moles per unit volume of air for total dissolved sulfur (i.e., other than $SO_4^{2-}$) in water of indicated pH at 20°C for the indicated air concentrations of $SO_2$ (Hales and Sutter, 1973; Dana, 1976).

where, again, $\tilde{F}_z$ is the $z$ component (positive, up) of the (turbulent and molecular) diffusive flux and $\tilde{d}_z$ is the positive $z$ component of any average drift velocity that the material might possess, other than the drift for the material attached to precipitation. The average in Eq. (4.36), denoted by the tilde, is the average over the high-frequency portion of the turbulence spectrum.

To develop Eq. (4.36) in a way that permits an evaluation of dry deposition, it is clearly necessary to specify $\tilde{F}_z$ and $\tilde{d}_z$; to do this, a number of approaches are available. For example, for the turbulence contribution to the flux, one can return to Eq. (4.5) and work with the defining correlation between fluctuations in the pollutant's air concentration and the vertical component of the wind field: $\widetilde{w'\chi'}$. This approach, however, is more exacting than the present state of the art can support for the prediction of average material fluxes to the ocean. For a parameterization of the turbulent flux, one can introduce the gradient-flux ansatz as in Eq. (4.6): $-\widetilde{w'\chi'} = K_z \partial\tilde{\chi}/\partial z$, where $K_z$ is the vertical component of the turbulent diffusivity. Some aspects of this approach will be used in the sequel. However, for now, a most illuminating analysis follows from a still coarser description. In this approximation, the mean concentration gradient is replaced by a difference in mean concentrations between two adjacent layers of the atmosphere, divided by the height difference, $\delta$, between where the mean concentrations are measured:

$$-\widetilde{w'\chi'} \simeq K_z \frac{\partial\tilde{\chi}}{\partial z} \simeq k_I(\tilde{\chi}_{I+1} - \tilde{\chi}_I), \qquad (4.37)$$

where $k_I = K_z/\delta_I$ and therefore $k_I$ is a net velocity at which the material is transferred by turbulence across the $I$th layer. As will be seen later, similar approximations can be introduced for transfer by processes other than turbulence. The major advantage of this simple approach is that, with it, it is relatively easy to identify the rate-limiting stage of the dry deposition process.

A. TRANSFER VELOCITIES

To utilize the parameterization shown on the right-hand side (RHS) of Eq. (4.37), it is first necessary to develop estimates for the transfer velocities within various atmospheric (and, similarly, oceanic) layers. To obtain these transfer velocities, consider the atmospheric layers shown in Figure 4.8. On the extreme RHS of Figure 4.8, the atmospheric layers are identified with letters A through D; in the second column from the RHS are shown approximate layer depths, and these lead to the transfer velocity estimates shown in the next column. Finally, the ranges of the velocity estimates are plotted as horizontal bars in the middle of Figure 4.8. The appropriate abscissa is at the top of the figure; the horizontal spread of the bars is derived by accounting for the variation in environmental conditions shown.

As explanatory notes to Figure 4.8 the following should be mentioned. Far from the interface and for the trace constituents of interest, the dominant transfer mechanism is almost invariably turbulence. In the constant flux layer, the mixing velocity is approximately $ku_*$, where $k$ is von Kármán's constant and $u_* = (\tau/\rho_a)^{1/2}$ is the friction velocity (in which $\tau$ is the shear stress and $\rho_a$ is the density of air); this result can most easily be obtained using $K_z \simeq k u_* z$ and $\delta = z$ in Eq. (4.37). The approximation $u_* \simeq 3\% \, \bar{u}$ is based on experimental data discussed later. For high-molecular-weight gases, the transfer velocity within the deposition layer is reduced by the factor $Sc^{2/3}$, where Sc is the Schmidt number, in an effort to correct for both the reduced layer thickness and the decrease in diffusivity, but this is only an approximation even for flow over a smooth, solid surface (Slinn *et al.*, 1977). Finally, the transfer velocity for particles includes the gravitational settling term, $v_g$, and any other slip velocity, $v_s$, such as might be caused by gradients in turbulent fluctuations or diffusiophoresis. The similar transfer velocities in the ocean, given in the lower half of Figure 4.8, and the dependence on the Henry's law constant, $H$, and effective solubility coefficient, $\alpha^*$, will be described in subsequent paragraphs.

B.   DRY DEPOSITION OF PERFECTLY ABSORBED GASES

A formalism is now developed that uses these transfer velocities and displays the rate-limiting stage of the dry deposition process. Consider, first, the transfer through the atmosphere of low-molecular-weight gases that are completely adsorbed by the ocean; as will be seen later, an example of this case is $SO_2$. Then the drift velocity $\tilde{d}_z$ of Eq. (4.36) can be ignored. If the remaining gradient terms (for turbulent and molecular diffusion) are approximated as in Eq. (4.37) then for steady-state conditions (and therefore continuous flux through all layers) the flux can be approximated by

$$-F_z = k_A(\tilde{\chi}_A - \tilde{\chi}_B) = k_B(\tilde{\chi}_B - \tilde{\chi}_C) = k_C(\tilde{\chi}_C - \tilde{\chi}_D) = k_D(\tilde{\chi}_D - \tilde{\chi}_0), \quad (4.38)$$

where $\tilde{\chi}_I$ is the mean gas concentration at the top of the *I*th layer. For this case of a perfectly absorbed gas, $\tilde{\chi}_0$ could be taken as zero, but retaining the symbol is convenient for the subsequent development. Alternatively, an overall atmospheric transfer velocity could be defined via

$$-F_z = k_a(\tilde{\chi}_A - \tilde{\chi}_0) \quad (4.39)$$

or, identically,

$$-F_z = k_a[(\tilde{\chi}_A - \tilde{\chi}_B) + (\tilde{\chi}_B - \tilde{\chi}_C) + (\tilde{\chi}_C - \tilde{\chi}_D) + (\tilde{\chi}_D - \tilde{\chi}_0)]. \quad (4.40)$$

The advantage of Eq. (4.40) is that if $(\tilde{\chi}_A - \tilde{\chi}_B)$ etc. are substituted from Eq. (4.38) into Eq. (4.40) and $F_z$ is canceled from the result, then we obtain

TRANSFER VELOCITY OR CONDUCTANCE, $k$ (cm s$^{-1}$)

| | | VELOCITY ESTIMATES | APPROXIMATE DEPTHS (m) | LAYER |
|---|---|---|---|---|
| ALOFT LAYER | DOWNDRAFTS / SUBSIDENCE / UNSTABLE / STABLE | $\approx \frac{K}{10 \text{ km}}$ | $10^4$ | A |
| BOUNDARY OR MIXED LAYER | | $\approx \frac{K}{1 \text{ km}}$ | $10^2 - 10^3$ | B |
| CONSTANT FLUX OR MECHANICAL TURBULENCE LAYER | $(u_*)_a \simeq 3\% \bar{u}$ | $\approx \ell (u_*)_a$ | $10^1 - 10^2$ | C |
| D≃ν | LOW MOLECULAR WEIGHT GASES | $\approx \ell (u_*)_a$ | $\frac{\nu}{\ell (u_*)_a} \left( \frac{\nu}{\ell u_*} \right)_a$ | D |
| D≪ν | Sc = ν/D | $\frac{\ell (u_*)_a}{(Sc)_a^{2/3}}$ | $\frac{\ell}{(Sc)_a^{1/3}}$ | |
| PARTICLES | DIFFUSION AND $\bar{u}$ = 10 m s$^{-1}$ | | $v_d = \left[ v_g + v_s + \frac{\ell u_*}{Sc_a^{2/3}} \right]_a$ | |

GAS PHASE (ATMOSPHERE)

$K$ (cm$^2$ s$^{-1}$) = $10^6$   $K = 10^5$   $10^4$   $10^3$

$\bar{u}$ = 10   1   0.1   m s$^{-1}$

eg., Sc = 30:

GRAVITATIONAL SETTLING : $a$ (μm) = 1 $a \simeq 1$ Å   $a = 0.1$ μm

$\bar{u}$ = 10 m s$^{-1}$   $\bar{u}$ = 1 m s$^{-1}$

RESUSPENSION

84

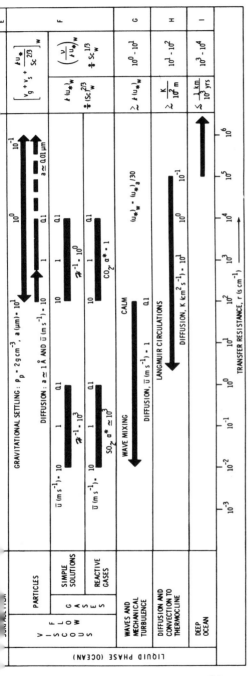

FIGURE 4.8 A multilayer model for dry deposition of particles and gases to the ocean. For specific conditions in each layer, specific points in each range can be identified; the layer in which the points lie farthest to the right in the figure is the layer that offers the greatest resistance to the dry deposition flux. The alphabetical symbols at the far right of the figure identify the symbols used in the formulas in the text.

85

$$\frac{1}{k_a} = \frac{1}{k_A} + \frac{1}{k_B} + \frac{1}{k_C} + \frac{1}{k_D}. \tag{4.41}$$

As a final step, if we set $r_I = k_I^{-1}$, then Eq. (4.41) becomes

$$r_a = r_A + r_B + r_C + r_D = \sum r_I, \tag{4.42}$$

which is the desired result since it sets the overall atmosphere "resistance" (the reciprocal of the overall transfer velocity, or "conductance," $k_a$) equal to the sum of the transfer resistances in the individual layers.

These transfer resistances, the inverses of the transfer velocities, are shown in Figure 4.8 on the abscissa at the bottom of the figure. The horizontal spread of each "resistance bar" in Figure 4.8 shows the range of resistances that "typically" occurs in the appropriate layer as environmental conditions range over the values shown. For specific conditions in each layer, specific points in each range can be identified. The point the farthest to the right in the figure identifies the layer that impedes the dry deposition flux with greatest resistance. Thus, if the resistance in layer $I$ is greatest, then transfer through the $I$th layer is the rate-limiting stage of the overall dry deposition process, and the overall atmospheric transfer rate, $k_a$, can be set as approximately equal to $k_I$. However, it is seen in Figure 4.8 that, at least for low-molecular-weight gases, the resistances in the atmospheric layers are typically of comparable magnitude (about 1 sec cm$^{-1}$), and unless there is a very stable layer aloft or the wind speeds are low there is no clearly defined rate-limiting stage; typically, then, for low-molecular-weight perfectly absorbed gases, $k_a \simeq 1$ cm sec$^{-1}$.

The transfer velocity for high-molecular-weight gases can be substantially smaller than 1 cm sec$^{-1}$. For example, for high-molecular-weight gases that are highly soluble in water (e.g., see Table 4.5 for solubility coefficients), the rate-limiting stage appears to be the atmosphere's viscous sublayer of Figure 4.8. If the transfer across this viscous sublayer is governed by Brownian diffusion, this then gives

$$k_a \simeq k_D \simeq k(u_*)_a/(\text{Sc})_a^{2/3}, \tag{4.43}$$

where $\text{Sc} = \nu/D$, $\nu$ is the kinematic viscosity of air, and $D$ is the molecular diffusion coefficient for the gas species in air. It must be emphasized, however, that this estimate is tentative until further research on the air–sea interface is performed. A discussion of specific research topics will be deferred until a later section.

### C. DRY DEPOSITION OF PARTICLES

For particles, the magnitude of the dry flux can be parameterized as

$$D_d = (K_z + D)\frac{\partial \tilde{\chi}}{\partial z} + (v_g + \tilde{v}_s)\,\tilde{\chi} + \widetilde{v_s'\chi'}, \qquad (4.44)$$

in which $v_g$ is the gravitational settling speed and $v_s$ is the (positive, down) component of any other slip velocity the particles might possess, e.g., caused by diffusiophoresis or gradients in turbulent fluctuations. As it stands, Eq. (4.44) cannot be conveniently recast into a form similar to Eq. (4.37) so as to display differences in concentrations. However, above the deposition layer, turbulent transport usually dominates the transfer of particles of interest, and, therefore, in these layers the other terms in Eq. (4.44) can usually be ignored. Then, in the deposition layer, where the turbulent fluctuations are damped, the downward flux can be parameterized (see Table 4.1) by the difference

$$D_d = v_d\tilde{\chi}(z_i) - v_r\tilde{C}(z_i) = v_d(\tilde{\chi}_D - \tilde{\chi}_*), \qquad (4.45)$$

where $v_d$ and $v_r$ are deposition and resuspension velocities, respectively, and $\tilde{\chi}_* = (v_r/v_d)\tilde{C}(z_i)$ is the average air concentration, which, when resuspension occurs, would be in balance with the pollutant's concentration in the ocean. For example, for NaCl sea-salt particles

$$\tilde{\chi}_* \simeq 1\,\frac{\mu g}{m^3}\left(\frac{\bar{u}}{1\ m\ sec^{-1}}\right)^n, \qquad (4.46)$$

where $n$ is about 3 and could be deduced more accurately from available data (e.g., see Junge, 1963). For $SO_4^{2-}$ particles, $\tilde{C}(z_i)$ is about a factor of 5 smaller than for Na; for Sr, $\tilde{C}(z_i)$ is about 3 orders of magnitude smaller; therefore, the $\chi$'s for these species would be correspondingly smaller (ignoring possible fractionation effects). With these approximations, the flux can be written as

$$D_d = k_A(\tilde{\chi}_A - \tilde{\chi}_B) = k_B(\tilde{\chi}_B - \tilde{\chi}_C) = k_C(\tilde{\chi}_C - \tilde{\chi}_D)$$

$$= v_d(\tilde{\chi}_D - \tilde{\chi}_*) = k_a(\tilde{\chi}_A - \tilde{\chi}_*). \qquad (4.47)$$

Proceeding as in the development of Eq. (4.41), Eq. (4.47) leads to

$$\frac{1}{k_a} = \frac{1}{k_A} + \frac{1}{k_B} + \frac{1}{k_C} + \frac{1}{v_d} . \tag{4.48}$$

In other words, the overall resistance is the sum of resistances in the individual layers, with $v_d^{-1}$ being the resistance to particle transfer across the deposition layer.

As can be seen from Figure 4.8, for particles (just as was the case for high-molecular-weight relatively soluble gases), most of the atmospheric transfer resistance occurs in the deposition layer; but for both types of atmospheric trace constituents, the resistances in this layer are not known at all well. For particles of radii $a$ in the range $0.001 \lesssim a \lesssim 0.1$ $\mu$m, the transfer velocity would be given by Eq. (4.43) for transfer to a smooth, solid surface, ignoring electrical and phoretic effects; in other words, these particles can effectively be considered as large gas molecules, and, consequently, Brownian diffusion dominates the transfer across the deposition layer. The resulting resistance bar is shown as the upper of the two parallel bars shown in the row labeled "particles" in Figure 4.8. Depending on the wind speed (i.e., on $u_*$), the transfer velocity would begin to increase again for particles with $a \simeq 0.1$ $\mu$m because of the greater particle mass; the resulting resistance bar is shown as the lower of the parallel bars, resistance decreasing to the left in Figure 4.8. The dotted bar to the right is included in Figure 4.8 to emphasize the importance of resuspension. However, it should be reiterated that these resistances are not known well; some recommended research will be mentioned in a later section.

For now, the best information available for particle deposition velocities is that obtained from water–wind-tunnel measurements; these data are shown in Figure 4.9. For particles smaller than 0.1 $\mu$m these results suggest that the increase in $v_d$ with increasing diffusivity given by a $Sc^{-2/3}$ dependence may be correct; however, waves, "sea spray," and induced motion of the water surface in the wind tunnel possibly did not adequately simulate conditions at sea. For particles larger than about 1 $\mu$m, it is seen that $v_d$ is substantially larger than the gravitational settling speed, $v_g$; indeed, it appears to be larger than the corresponding inertial impaction contribution to $v_d$ for deposition to a smooth surface (Sehmel, 1973; Caporaloni et al., 1975; Slinn, 1976d). Possibly this is caused by an increase in inertial impaction on waves; if so, then failure to simulate ocean waves in the wind tunnel should limit

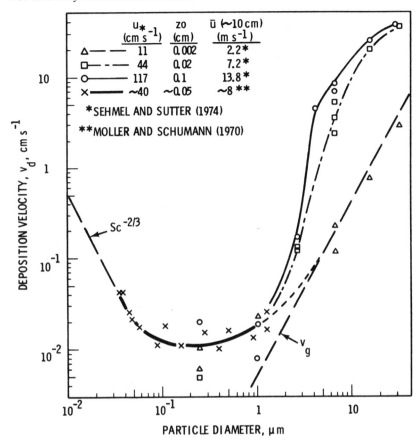

FIGURE 4.9 Deposition velocity as a function of particle size as measured by Sehmel and Sutter (1974) for particles of density 1.5 g cm$^{-3}$ depositing on a water surface in a wind tunnel and by Möller and Schumann (1970) for similar conditions using polystyrene latex spheres, sodium chloride particles, and purified-water residual nuclei.

generalizations from the data. Finally, for particle diameters in the range 0.1 to 1 $\mu$m, the deposition velocity shown in Figure 4.9 appears to be the result of two processes: gravitational settling and Brownian diffusion. However, as will be discussed in more detail later, for conditions over an evaporating ocean, diffusiophoresis would be expected to decrease the deposition rate while thermophoresis would increase it.

D.  DRY DEPOSITION OF GASES

For the case of gases that are not perfectly absorbed by the ocean, it is imperative to account for their flux back to the atmosphere. To simplify the analysis in this case, it is convenient to reformulate the problem and develop only a two-layer model. In the atmospheric layer, all transfer rates are lumped into a single overall gas-phase transfer rate, $k_g$, between the air–sea interface and a convenient height in the atmosphere where a bulk concentration, $\tilde{\chi}_b$, is measured. Then

$$D_d = k_g(\tilde{\chi}_b - \tilde{\chi}_i). \qquad (4.49)$$

In the ocean layer, similarly,

$$D_d = k_l(\tilde{C}_i - \tilde{C}_b), \qquad (4.50)$$

where $k_l$ is the transfer velocity in the liquid between the surface and the depth for convenient measurement of the bulk concentration, $\tilde{C}_b$. The two expressions for the dry flux, Eqs. (4.49) and (4.50), can now be equated and lead to an estimate for the total transfer rate. Before describing this, however, it is useful to discuss transfer processes in the ocean.

In the above discussion, the emphasis has been on developing parameterizations for transfer processes in the atmosphere, starting from the continuity equation, Eq. (4.1). Clearly, though, a similar parameterization for Eq. (4.2) could have been described and would lead to an overall transfer velocity in the ocean

$$\frac{1}{k_o} = \frac{1}{k_F} + \frac{1}{k_G} + \frac{1}{k_H} + \frac{1}{k_I}, \qquad (4.51)$$

where the subscripts $F$ through $I$ label the layers in the ocean shown in Figure 4.8. There, too, are shown estimates for the transfer velocities through the individual layers, estimated in a manner similar to the atmospheric case. In particular, it is noted that, by equating stresses at the interface, $(pu_*^2)_w = (pu_*^2)_a$, which provides an estimate for $(u_*)_w$. Also, it should be noted that because the molecular diffusion of gases in water is slow (typically $D \sim 1$ cm$^2$ day$^{-1}$), whereas $\nu_w \simeq 0.01$ cm$^2$ sec$^{-1}$, the Schmidt number is typically very much larger than unity. The appropriate $k_l$ to use in Eq. (4.50) depends on the depth to where the "bulk" concentration, $\tilde{C}_b$, is measured, usually a few meters below the surface.

We now return to Eqs. (4.49) and (4.50) for the case of gases that form simple solutions in the ocean, and, consequently where Henry's law prevails at the interface:

$$\tilde{\chi}_i = H\tilde{C}_i. \tag{4.52}$$

If the flux is also written as

$$D_d = k_t(\tilde{\chi}_b - \tilde{\chi}_*), \tag{4.53}$$

where $k_t$ is an overall or total transfer velocity and $\tilde{\chi}_* = H\tilde{C}_b$ is the air concentration that would be in equilibrium with the existing bulk concentration of the gas in the ocean, then algebra similar to that used earlier leads to

$$D_d = k_t\left[(\tilde{\chi}_b - \tilde{\chi}_i) + H(\tilde{C}_i - \tilde{C}_b)\right], \tag{4.54}$$

or

$$\frac{1}{k_t} = \frac{1}{k_g} + \frac{H}{k_l}, \tag{4.55}$$

which relates the total transfer velocity to the transfer velocities in the two media. Incidentally, instead of Eq. (4.53), sometimes it is convenient to write the flux in terms of liquid-phase concentrations; viz.,

$$D_d = K_t(\tilde{\chi}_b/H - \tilde{C}_b), \tag{4.56}$$

from which it is seen that $K_t = Hk_t$.

In Eq. (4.55) the total resistance to transfer, $r_t = k_t^{-1}$, appears as the sum of a gas-phase resistance $r_g = k_g^{-1}$ and a liquid-phase resistance $r_l = Hk_l^{-1}$:

$$r_t = r_g + r_l. \tag{4.57}$$

Figure 4.8 shows that for solubility coefficient $\alpha = H^{-1} \simeq 10^0$, the liquid-phase transfer resistance dominates; Table 4.5 shows that many low-molecular-weight gases fall in this category. Some pesticides and herbicides fall in the intermediate range, $10^1 \lesssim H^{-1} \lesssim 10^2$, and the resistances in the atmosphere and the ocean are comparable in magnitude (Munnich, 1971). For highly soluble gases or gases that react in the ocean to form relatively nonvolatile products, their transfer is gas-phase controlled.

Actually, for the case of reactive gases such as $SO_2$, it is necessary to strain the formalism somewhat, in an attempt to account for gas "transfer" to reaction products. For example, for a first-order reaction with reaction rate constant $\beta$, averaging Eq. (4.2) over high-frequency turbulence and accounting only for vertical diffusion leads to

$$\frac{\partial \tilde{C}}{\partial t} + \tilde{v}_t \cdot \nabla \tilde{C} = \frac{\partial}{\partial z} \left[ (K_{zl} + D_l) \frac{\partial \tilde{C}}{\partial z} \right] - \beta \tilde{C}. \qquad (4.58)$$

One way to modify the formalism to account for this case is first to identify the rate-limiting stage of the transfer process in the ocean. If this is the viscous sublayer, then from Eq. (4.58) an effective flux "through" this layer can be identified as

$$-F_z = k_F (\tilde{C}_F - \tilde{C}_G) + \beta \delta_d (\tilde{C}_F - \tilde{C}_G), \qquad (4.59)$$

where $k_F = [K_{zl} + D_l]/\delta_d$, in which $\delta_d$ is the thickness of the layer, and where the introduction of a nonzero $C_G$ in the last term of Eq. (4.59) cannot be justified except on the basis of mathematical convenience. The convenience is that Eq. (4.59) then leads to an effective transfer velocity

$$k_F^* = k_F + \beta \delta_d, \qquad (4.60)$$

i.e., enhanced by the chemical reactivity of the gas. In turn, if Eq. (4.60) is used in Eq. (4.55) the total transfer velocity can be written as

$$\frac{1}{k_t} = \frac{1}{k_g} + \frac{H}{\alpha^* k_l}, \qquad (4.61)$$

where $\alpha^*$ is an effective enhancement of the solubility of the gas in water. As is shown in Figure 4.8, $\alpha^* \simeq 1$ for $CO_2$ and $\alpha^* \simeq 10^3$ for $SO_2$ (Liss and Slater, 1974). These considerations lead to the prediction that the transfer of highly soluble or reactive gases to the upper layers of the oceans is controlled by transfer through the atmosphere. The restrictive comment about the ocean's upper layers follows from the presence of the bottom resistance bar in Figure 4.8, which shows that transfer to the deep ocean (and ultimately to the ocean floor) is orders of magnitude slower than other transfer processes and therefore is the rate-limiting step for the final deposition of most airborne material. (We exclude from consideration the possible modifying effects of biological processes on deposition rates in the ocean.)

### E. SUMMARY AND COMMENTS

This section is best summarized by Figure 4.8. It may be useful to emphasize, however, that this development has restricted value since, clearly, it is based on crude parameterizations of the governing equation and crude approximations for the transfer velocities. In contrast, if short-term estimates are desired (e.g., to predict the fate of a specific pollutant or tracer release or to support limited-duration field experiments) then it probably would be necessary to retain the gradient-flux parameterization of the turbulence flux or to return to the correlation $\widetilde{w'\chi'}$, the choice depending on the accuracy with which other factors are known. On the other hand, for long-term average estimates of dry flux of pollutants to the ocean, the type of parameterization developed in this section is probably the best available. Nevertheless, it is clear that a number of studies should be undertaken to strengthen the parameterizations summarized in Figure 4.8; suggestions for such studies will be made in the next section.

## V. SOME CURRENT AND SUGGESTED RESEARCH

In the previous sections, a framework has been presented for the prediction of atmospheric trace constituent fluxes through the air–sea interface. In the presentation, a number of weaknesses in the framework have been noted, but little attention was paid to how the framework might be strengthened. In this section, the emphasis will be on current and suggested future research that might strengthen prediction capabilities. The ordering of topics will generally follow the order in which the topics appeared in previous sections.

### A. RADIONUCLIDE MEASUREMENTS

Measurements and analyses of natural and anthropogenic radionuclide concentrations can provide valuable information on mixing, or inter-reservoir transfer, and on wet and dry removal processes. For the analyses of many of these data, it is assumed that nongaseous radionuclides rapidly attach to atmospheric aerosol particles, and, therefore, from these data inferences can be made about the behavior of atmospheric aerosols. Because this is a fundamental assumption, further studies of this attachment process and its variability would be welcome, e.g., along the lines developed by Lassen *et al.* (1960; 1961) and Brock (1970). To illustrate the importance of this type of analysis, it is noted that, in their recent critical review, Martell and Moore (1974) conclude that the best estimates of tropospheric aerosol residence times are those based on $^{210}$Bi to $^{210}$Pb activity ratios as well as those based on the average $^{210}$Pb concentrations versus altitude in the troposphere over central continental areas. Each of these methods indicates an average tropospheric aerosol residence time of less than one week. However, as reported by these authors, size distribution measurements have indicated that 90 percent of the $^{210}$Pb and $^{210}$Bi activity can be associated with particles smaller than 0.3-$\mu$m diameter (Gillette *et al.*, 1972), and the presentation in the previous sections would suggest that such particles would have a relatively longer residence time than would larger particles. Consequently, to increase the utility of radionuclide data, further research to characterize the radionuclides and their host particles would be welcome.

Extensive data are available for wet and dry fallout of radionuclides, especially for fission products deposited on land. Technological breakthroughs are required to measure dry deposition rates to the ocean and, in some cases, even to the land (e.g., for $^{210}$Pb and $^{210}$Bi). Some data are, however, available for the sea surface for fission products (Schumann, 1975). The ratio of dry to wet deposition for $^{90}$Sr was found to be about 0.2 and independent of latitude in 10° S to 45° N; similar values were found for $^{210}$Pb. These values, which represent upper limits, are similar to, or even lower than, the values observed over land for fission products and trace elements (Cambray *et al.*, 1973; Cawse, 1974). However, to generalize from these results in an attempt to infer the relative importance of wet and dry removal to the oceans would ignore the different source functions for these materials, the differences in the particle sizes to which the radionuclides are attached, and the differences for dry deposition to mechanical collectors as opposed to natural surfaces.

The observations by Bowen and his colleagues of higher integrated $^{90}$Sr activity in the oceans have led to speculations of possible mechanisms that might be responsible for higher fallout over oceans (Bowen and Sugihara, 1960; Volchok *et al.*, 1971; Volchok, 1974). From the information presented in the previous paragraph, some authors have concluded that dry deposition can be ruled out as an important contributing factor. Several other factors such as higher rainfall over ocean and land runoff have been considered, but at present this problem has not been resolved.

From the above paragraphs one can see that, whereas it is a relatively simple matter to estimate the total radionuclide fallout at a specific place, it is nevertheless difficult to extrapolate from available measurements to estimate the global values for fallout, over oceans or land. As a further example, it is now well documented that for several land sites, the specific activity in rains is relatively independent of total rainfall. Thus, Crooks *et al.* (1960) found that for five stations in the United Kingdom where annual rainfall varied between 50 and 320 cm, the specific activity of $^{90}$Sr remained between 5.7 and 6.5 pCi/liter. However, the results for some coastal stations show a very good inverse correlation (Lal *et al.*, 1976), and, in some cases, a weak positive dependence has been found, e.g., for $^{210}$Pb data in Hokkaido, Japan (Fukuda and Tsunogai, 1975). It can therefore be seen that considerable uncertainty remains in estimating both the global-average values of fallout or relative land–ocean values.

In order to improve prediction capabilities, better and more complete data are needed. Especially useful would be more information on air concentrations, attachment processes, and wet–dry fallout of the radionuclides $^{7}$Be, $^{210}$Po, $^{210}$Bi, and $^{210}$Pb and trace elements such as Hg, As, Pb, and Zn. The fallout studies should seek relationships between the concentrations in air and in precipitation for oceanic, coastal, and inland stations. In view of the inadequacy of mechanical dry deposition samplers (see Chapter 10), the obvious solution should not be overlooked—to study in detail the actual flux of radionuclides to the ocean itself and the fate of the radionuclides in the ocean.

### B. PRECIPITATION SCAVENGING STUDIES

A good survey of current activities in wet removal research is available in the conference proceedings *Precipitation Scavenging—1974* (Beadle and Semonin, 1976). Much of the current research is focused on evaluating the particle collection efficiency of various types of precipitation (for a review, see Slinn, 1976c), determining the deposition patterns of tracers released into precipitating storms (for a review, see Gatz, 1976b), and determining the deposition downwind of specific sources. Rain scavenging of gases has been extensively investigated by Hales and his co-workers (e.g., see Dana *et al.*, 1975). Earlier precipitation scavenging conference proceedings are those edited by Engelmann and Slinn (1970) and Styra *et al.* (1970).

To assist in quantifying the wet flux of pollutants to the oceans, the washout ratio measurements for a number of pollutants by Gatz (1976a), Peirson *et al.* (1974), and Cawse (1974) are especially valuable. Some of their data were used earlier in this chapter to compile Table 4.4. The theoretical analyses of washout ratios outlined earlier in this chapter may be able to explain many of the observed variations of washout ratios as a function of particle size, rain intensity, drop size, and storm type and with the vertical profiles of the pollution. Needed are measurements of washout ratios over the oceans (in fact, at any location) coupled with vertical profiles of the trace constituents ingested by the storms and with details about the precipitation characteristics.

Our knowledge of precipitation scavenging would be greatly improved by a coordinated series of projects in which budget studies are made on progressively more complicated storm systems. It is envisaged that these projects would start with simple wave and cap clouds, progress to orographic and cumulus clouds, and culminate in a study of cumulonimbus and frontal storms. The difference between the inflow and outflow of airborne material (including water vapor) should be compared with the removal by precipitation. Budgets should be performed for water vapor, trace elements, simple compounds, ions, and inorganic and organic gases. Particle size distribution measurements in the inflow and outflow air could yield information on the modification of the aerosols by clouds, and similar studies could yield information about clouds as sources and sinks of cloud droplet and ice-crystal nuclei. There are indications that studies such as these may soon be performed in the northeastern United States as a part of regional air-pollution studies.

Also required is more statistical information about storms—average duration, total precipitation, cloud-water removal efficiencies, and frequency of occurrence. Over the oceans, even the total rainfall is uncertain: $^{90}$Sr fallout may represent the best available rain gauge! Marwitz (1972) and Browning and Foote (1976) have discussed the dramatic ($\sim$ order of magnitude) decrease in precipitation efficiency of convective storms as the wind directional shear from cloud base to cloud top increases by an order of magnitude, but there are few data available for the cloud-water removal efficiency for frontal storms. In addition to field projects described in the previous paragraph, further development of remote sensing from satellites might yield valuable statistical information on storms and should be encouraged. Probably much information could also be obtained from existing global circulation models.

C. MICROMETEOROLOGIC FACTORS INFLUENCING DRY
DEPOSITION

Since the turbulent motions that generally control the dry transfer of
atmospheric pollution to the vicinity of the sea surface are precisely
those that contribute to the vertical fluxes of sensible heat, moisture,
and momentum, studies of these latter fluxes can be used to infer
information about the turbulent transfer of pollutants. Considerable
experimental effort has resulted in a fairly good understanding of the
oceanic case, at least in wind speeds below about 15 m sec$^{-1}$. Present
knowledge can be summarized by the observations that (a) a surface
layer in which fluxes are rapidly equilibrated ($\sim$ minutes) extends up to
20–50 m typically (see, however, Deacon, 1973), (b) empirical descriptions, obtained over land, of flux–gradient relationships apply equally
well over water, and (c) the wind drag at the surface $\tau$ is well described
by a simple drag coefficient relationship involving the air density $\rho$ and
the mean wind speed $\bar{u}$:

$$\tau = C_d \rho \bar{u}^2, \tag{4.62}$$

where $C_d$ is found to be approximately $1.3 \times 10^{-3}$ in moderate winds
and in near-neutral stratification and approximately $1.4 \times 10^{-3}$ for
slightly unstable conditions that are typical at sea. However, variations
in $C_d$ with wind speed have been suggested, although, as a dimensionless quantity, $C_d$ should not depend only on the mean wind but possibly
on some dimensionless characteristic of the wave spectrum. A dependence of $C_d$ on $c/u_*$, where $c$ is the phase speed of the significant waves
and $u_*$ is the friction velocity, has been suggested, although such a
dependence should be expected only for an equilibrium-wave spectrum. The experimental evidence is perhaps fortuitous since relations
between $C_d = u_*^2/\bar{u}^2$ and $c/u_*$ are influenced by experimental errors in
$u_*$.

The friction velocity $u_* = (\tau/\rho)^{1/2}$ is a convenient and useful measure
of the turbulence intensity. For cases for which the wind speed is
measured directly (usually at a height of 10 m), the simple proportionality

$$u_* = 0.037\bar{u}_{10} \tag{4.63}$$

is compatible with the drag coefficient given in Eq. (4.62) for typical
conditions at sea. For those situations in which a direct measurement
of the velocity is not available, it is possible to estimate $u_*$ from the

geostrophic wind speed $u_g$ based on surface pressure gradients (Deacon, 1973; Hasse, 1974)

$$u_* \simeq 0.025 u_g. \qquad (4.64)$$

Errors involved in the evaluation of $u_*$ from these relatively simple relationships are likely to be of the order of $\pm 20$ percent. Similar errors can arise from the neglect of atmospheric stability; corrections for stability can be derived from the relationships recommended by Dyer (1974) and Hasse (1974). However, it is unlikely that stability corrections are important for contaminant profiles if the controlling resistance to dry deposition is associated with heights below a few decimeters, provided that the proper $u_*$ is used.

A number of reports have recently focused on strengthening the electrical analogy for trace-constituent transport through the constant flux layer. A later subsection will address the case of particle transport; for gases, the analogy can be developed by integrating the gradient-flux ansatz

$$-F \equiv D_d = K(z) \frac{\partial \chi}{\partial z}, \qquad (4.65)$$

where the diffusivity $K$ contains both turbulent and molecular contributions. In general, if the dry flux is constant, then integration of Eq. (4.65) from the interface $z = z_i$ leads to

$$D_d = k_g \left[ \chi(z) - \chi(z_i) \right], \qquad (4.66)$$

where

$$k_g^{-1} \equiv r_g = \int_{z_i}^{z} dz' / K(z'). \qquad (4.67)$$

If it is assumed that the dominant contribution to $K$ above the aerodynamic roughness height $z_0$ is from turbulent diffusion, then in the constant flux layer and for neutral density stratification we have

$$K \simeq k u_* z, \qquad z \geq z_0, \qquad (4.68)$$

in which $k$ is von Kármán's constant. Substituting Eq. (4.68) into Eq. (4.67) leads to

$$r_g = \int_{z_i}^{z_0} \frac{dz'}{K(z)} + (k u_*)^{-1} \ln \frac{z}{z_0} \equiv r_m + r_a, \qquad (4.69)$$

in which an aerodynamic resistance, $r_a = (ku_*)^{-1} \ln(z/z_o) \equiv \bar{u}/u_*^2$, and an atmospheric surface-layer resistance where molecular diffusion dominates, $r_m$, have been identified separately. Many expressions for this surface-layer resistance have been suggested in an attempt to account correctly for pollutant-dependent heights at which molecular diffusion dominates and to compensate for the lack of an equivalent to pressure (or form) drag (Owen and Thompson, 1963; Dipprey and Sabersky, 1963; Sheriff and Gumley, 1966; Dawson and Trass, 1972; Garratt and Hicks, 1973; Yaglom and Kader, 1974; Brutsaert, 1975; Roth, 1975). Generally these expressions are of the form

$$u_* r_m \equiv B^{-1} = a \; \text{Re}_*^{\,b} \; \text{Sc}^c - d , \qquad (4.70)$$

where $a$, $b$, $c$, and $d$ are constants, $\text{Re}_* = u_* z_o/\nu$ is the surface Reynolds number and $\text{Sc} = \nu/D$ is the Schmidt number. For example, Dipprey and Sabersky (1963) gave $a = 10.25$, $b = 0.2$, $c = 0.44$, and $d = 8.5$. If this formalism is adequate for gases, then it should also apply for particles smaller than about 0.1 $\mu$m since, typically, inertial effects for such particles are negligible compared to their molecular (or Brownian) diffusion. However, it should be noted that expressions such as Eq. (4.70) have never been advanced specifically for the case of air–sea transfer, and, therefore, further research (e.g., see Shepherd, 1974) to test their adequacy in this case would be appropriate.

One of the difficulties in the application of these formulas to the case of dry deposition to the oceans is that the roughness length has no direct physical meaning at sea, but, rather, $z_o$ is a function of the sea state and the relative speed of the wind and waves. For example, in a recent review paper, Gifford (1976) quoted the suggestion by Kitaigorodskii (1970) that the "equivalent sand roughness" of the sea, $h_s$, depends on $u_*$ (or $\bar{u}$) relative to the peak phase velocity of gravity waves, $c_o$, according to

$$h_s = \begin{cases} \sigma, & u_* >> c_o & (4.71a) \\[2mm] 0.38 \, u_*^2/g, & u_* << c_o & (4.71b) \end{cases}$$

where $\sigma$ is the rms wave height and $g$ is the acceleration due to gravity. Thus during early stages of wave development, with $u_* >> c_o$, Eq. (4.71) suggests that the waves behave essentially as immobile roughness elements; later in the wave development, $u_* << c_o$ and Eq. (4.71) suggests that, aerodynamically, oceans are typically quite smooth ($h_s \lesssim 1$ cm). However, Eq. (4.71b) and similarly the Charnock relation $z_o \sim u_*^2/g$, although dimensionally correct, fail to represent the important role of capillary waves on the uptake of stress; that is, one would expect the surface tension, $\gamma$, to appear in a dimensionless correlation. Brocks and Krügermeyer (1972) have shown that $z_o \sim u_*^2/g$ is not correct under neutral conditions. In general, it must be concluded that the physics of stress at the sea surface are still uncertain.

Further, it is not just the roughness length that is uncertain but also the mean wind profile itself. Thus, although it has been found that atmospheric turbulence spectra above the waves in general do not show strong influences of wave motion, wind profiles in the trade wind regime were distorted below the wave heights in the sense that $\partial \bar{u}/\partial z$ was increased (Dunckel et al., 1974). This probably is a feature typical for swell only (that is, peak wave speed $c_o$ greater than wind speed). An interpretation is that waves take stress from, or feed stress to, the atmosphere depending on the relative speeds $c_o$ and $\bar{u}$. The slope of the profile should reflect the fraction of stress transported by turbulence in contrast to momentum uptake via pressure forces. The stress uptake by waves as a fraction of total stress is between 20 and 80 percent (Dobson, 1971; Hasselmann et al., 1973; Snyder, 1974). Considering that the wave field initially adjusts fairly rapidly (of the order of an hour) to changes in the wind field and that the drag coefficient seems to be fairly independent of the sea state, it seems reasonable to assume that the fraction going into waves will be at the lower margin (20 percent) in most cases, except shortly after rapid change. The consequences for material transfer would be proportional, in that a reduction of turbulence intensity would cause a corresponding *reduction* in turbulent transfer of pollutants; on the other hand, with momentum uptake by pressure on waves, there could be corresponding *increases* in inertial impaction of particles and convective transport of gases.

### D. THE AIR–SEA INTERFACE AND SURFACE-LAYER MIXING

The analyses in earlier sections suggest, however, that the most important details to study may be those of the air–sea interface itself and of mixing in the oceans. Thus, it was seen from simple analyses that in many cases dry transfer would be controlled in the viscous sublayers at the interface or by mixing in the top layers of the ocean (cf., Figure 4.8). It is, therefore, important to learn more about these layers, especially to determine if mechanisms exist to "short-circuit" the high transfer resistances predicted by simple theories. Here a few comments are made about relevant facets of the interface and mixing in the ocean's surface layer.

The momentum (or viscous-dominated) boundary layers themselves are probably not strongly influenced by waves and the slip of the surface. Typically the slip of the sea surface is a few percent of $\bar{u}_{10}$. Gravity waves are not expected to influence strongly the aqueous interfacial layer because their wavelength is large compared with the characteristic viscous length; capillary waves may be more important because of their smaller size. Waves may affect transports by their periodic stretching and compressing of the interfacial layer. If this effect is proportional to the change in surface area, then variations in thickness of about 15 percent could be expected (see below). However, although the momentum boundary layers may not vary significantly, there could be a significant alteration of the deposition-layer thicknesses and of the corresponding flux through these layers of high-molecular-weight gases and particles whose transfer is controlled by Brownian diffusion. Analysis of this problem [e.g., along the lines developed by Zimin (1964) and Slinn (1976c) for the similar problem of transport to raindrops with internal circulation] is desirable.

The physics of waves and wave breaking certainly should be investigated from the point of view of pollutant transport. The increase in surface area by waves is modest since wave steepness is limited. A typical steepness of longer gravity waves is 1/17, and if to this are added capillary waves with a steepness of, say, 1/7, then the increase in surface area would be about 15 percent. Cox and Munk (1955) give for the relative increase in surface area

$$\frac{\Delta A}{A} = 0.0015 + 0.0026\,\bar{u} \pm 0.002, \qquad (4.72)$$

with the masthead velocity, $\bar{u}$, given in m sec$^{-1}$. Waves begin to break at Beaufort 3, and spray blown from the crest of breaking waves starts to become noticeable at Beaufort 6. With higher wind speeds, the amount of spray in the air increases rapidly; at hurricane speeds it is said that the sea surface is no longer defined: the air is filled with spray. The increase of effective surface area with spray, the capture of aerosol particles by spray drops, the trapping of gas by breaking waves—these are some of the processes that deserve further investigation since all would effectively bypass the slow transport across the interfacial layers.

In contrast, interfacial resistance can be increased by surface films, a subject recently reviewed by Liss (1975; 1976). Natural and artificial surface films can potentially influence gas exchange either directly, i.e., by acting as an additional barrier to transfer, or indirectly by affecting some interfacial property that is important for gas transfer, e.g., by damping capillary waves. However, any effect will be severely reduced if the film material does not form a continuous layer at the interface. It has been argued (Garrett, 1972; Liss, 1975, 1976) that because the material found at the sea surface is generally of very mixed chemical composition, has low *in situ* film pressure, and can be considerably compressed before exhibiting any appreciable rise in film pressure, it is unlikely to form a continuous layer and so will not play a significant role in retarding gas exchange. Exceptions may occur in coastal and other areas of high biological production, in the vicinity of oil spills, or, possibly, in shipping lanes. It is further noted that although no systematic census is known to us, it is usually reported that oil slicks at sea break up at wind speeds of 3 m sec$^{-1}$ and higher. The absence of capillary waves (which become damped when a continuous film is present) is probably a sensitive indicator of areas where exchange retardation might be significant.

Mixing within the oceans is complicated and deserves further study. Near the interface, the turbulence intensity in the water is determined by the stress transmitted past the interface and by the velocity profile. Compared with a layer of, say, 1-cm depth at the interface, the orbital motions of most wave components decay on a larger depth scale (of the order of $\lambda/2\pi$, where $\lambda$ is the wavelength) and thus do not contribute to the velocity profile adjacent to the interface. For such a small layer, the velocity profile may be modeled as a logarithmic (constant stress) profile assuming the stress in the water to be equal to the part of the total stress transferred to the interface as shearing stress (compared with pressure forces on waves). Logarithmic profiles have been reported by Shemdin and Lai (1970) in the water layer of a wind–water tunnel. However, it is difficult to infer what the fraction of momentum taken by the waves will do for the turbulent transports in the near-surface water layers. If the wave energy is fed by wave–wave interactions to higher frequencies, then the energy taken up by longer waves would also be available for turbulence generation. This argument is based on the assumption that dissipation occurs mainly in the thin layers of orbital motions of the small waves; if it is correct, then small errors would be made in estimating pollution fluxes by assuming the total stress to pass through the interface as tangential stress. Mixing to deeper layers may be dominated by buoyancy (cooling of the surface by evaporation and effective net radiation enhances mixing; heating of the surface by strong insolation is rarely strong enough to produce stable stratification and inhibit mixing), but certainly ocean currents contribute to vertical mixing. The ultimate deposition of materials to the ocean floor depends, of course, on mixing past the thermocline and in the deep ocean, a subject about which the authors have limited knowledge.

### E. GAS EXCHANGE ACROSS THE AIR–SEA INTERFACE

Some aspects of the exchange processes requiring further investigation to strengthen the prediction capabilities for gas transfer will be mentioned in this subsection. Earlier, in Eqs. (4.53) and (4.56), the exchange flux was written in terms of an overall transfer velocity expressed on a gas $(k_t)$ or liquid $(K_t)$ phase concentration basis, with $K_t = Hk_t$. In turn, the overall transfer velocity is

$$\frac{1}{k_t} = \frac{1}{k_g} + \frac{H}{\alpha^* k_l}, \tag{4.73}$$

where the first term represents the resistance presented by the gas phase $r_g = k_g^{-1}$ and the second term, the resistance from the liquid phase, $r_l$. In Eq. (4.73), $\alpha^*$ is an effective enhancement of the gas solubility coefficient caused by chemical reactivity. In this subsection, current research and research needed to determine $r_g$ and $r_l$ will be discussed.

It is, of course, true that measurements of gas transfer will yield only the total resistance, $k_t^{-1}$ or $K_t^{-1}$. However, as is the case with the evaporation–condensation of any appreciably pure liquid, the aqueous-phase resistance should be negligible for water molecules crossing the air–sea interface, i.e., $r_l = 0$ (Whitney and Vivian, 1949). Consequently, by measuring the exchange of water molecules, along with the transfer of other gases, it is possible to divide the overall resistance to exchange into its gas- and liquid-phase components. A mean value for $k_g$ for $H_2O$ over the oceans is approximately 3000 cm h$^{-1}$ or 0.8 cm sec$^{-1}$ (Schooley, 1969). Direct measurements using the radon deficiency method (Broecker, 1965) together with results from studies of natural and bomb-produced $^{14}CO_2$ (Broecker and Peng, 1974) yield a value for $K_t$ of about 20 cm h$^{-1}$; as will be seen later, these results essentially yield $k_l \simeq 20$ cm h$^{-1}$.

For gases that are chemically reactive in water, $\alpha^*$ will be greater than 1 and the liquid-phase transfer velocity will be higher than for gases that are inert in the aqueous phase. Exchange enhancement has been observed in the laboratory for $CO_2$ and $SO_2$ (Hoover and Berkshire, 1969; Liss, 1973; Brimblecombe and Spedding, 1972; Slater, 1974), and a number of equations have been proposed that are reasonably successful in accounting for the magnitude of the effect (Hoover and Berkshire, 1969; Quinn and Otto, 1971; Emerson, 1975). The equations predict that under typical marine conditions the value of $\alpha^*$ for $CO_2$ is very close to unity (1.02–1.03) so that exchange enhancement due to chemical reactivity is only a few percent. In contrast, under similar conditions, values of $\alpha^*$ for $SO_2$ can be several thousand, which means that chemical reaction greatly increases the rate of exchange for this gas. The very large difference between $\alpha^*$ for $CO_2$ versus $SO_2$ arises because the rate of hydration for $SO_2$ ($3.4 \times 10^6$ $\text{sec}^{-1}$) is about $10^8$ times faster than the hydration rate for $CO_2$ ($3 \times 10^{-2}$ $\text{sec}^{-1}$).

Using the values for $k_g$ and $K_l$ given above, together with data for $H$ and calculated values for $\alpha^*$, Eq. (4.73) can be used to split the overall resistance to transfer for any gas into its gas- and liquid-phase components. It is found for gases of low solubility that are chemically unreactive in seawater ($\alpha^* \simeq 1.0$) that $r_l >> r_g$ (e.g., $N_2$, $O_2$, $CO_2$, $CH_4$, $N_2O$, and noble gases). In the case of gases of high solubility, rapid aqueous-phase chemistry, or both, $r_g >> r_l$ (e.g., $H_2O$, $SO_2$, $NO_2$, $NH_3$, $HCl$, and $HF$). These transfer velocities are average values for the whole ocean surface. They are satisfactory for calculating overall gas fluxes and have been used, along with average values for air–sea concentration differences, to obtain fluxes of the following gases across the sea surface: $SO_2$, $N_2O$, $CO$, $CH_4$, $CCl_4$, $CCl_3F$, $CH_3I$, and $(CH_3)_2S$ (Liss and Slater, 1974).

In order to refine these flux calculations, it is necessary to have a better knowledge of the concentrations of the gases of interest in seawater and in the marine atmosphere as well as more accurate values for the gas- and liquid-phase transfer velocities. Since many of the gases of interest occur at very low concentrations (down to 1 part in $10^{12}$ by volume), their determination in seawater and marine air must be carried out with great care. The questionable quality and small number of measurements of the air–sea concentration differences are probably the most important factors now limiting the accuracy of these calculations. However, for quite a number of gases, the methodology for making such measurements is available so that there is no major technological barrier to obtaining concentration data.

In order to obtain better values for $k_g$, a considerable amount of micrometeorological knowledge is available, although, as discussed in the previous subsections, further research is required. Micrometeorological techniques for the measurement of water-vapor fluxes, such as eddy correlation and the profile method, may be applicable to the direct determination of fluxes of other gases whose air–water exchange is controlled by gas-phase processes. However, profile methods cannot be used to find the flux of gases for which the principal resistance to transfer is in the water phase. For these gases, most of the concentration change will take place across the larger resistance, so that vertical gradients in the air will be very small.

When it comes to estimating values for the liquid-phase transfer velocity, very few *in situ* techniques are available. The radon deficiency method (Broecker, 1965), which involves measurement of the decrease in the $^{222}$Rn/$^{226}$Ra ratio in near surface waters due to loss of Rn to the atmosphere, is presently the most useful. Although the method gives an overall transfer velocity [viz., $K_l$, see Eq. (4.56)], nevertheless, because for Rn $r_l >> r_g$, the values obtained will, for all practical purposes, be equal to the transfer velocity of the liquid phase ($k_l$).

In contrast to $k_g$, which from both theoretical (Hicks and Liss, 1976) and experimental studies (Liss, 1973) increases linearly with wind velocity, laboratory wind-tunnel experiments indicate that $k_l$ increases approximately as the square of the wind speed (Downing and Truesdale, 1955; Kanwisher, 1963; Liss, 1973). Although some recent results (W. S. Broecker, 1976), obtained using the radon deficiency method on the GEOSECS cruises in the Atlantic, do show an increase in $k_l$ with wind speed, the rate of increase is considerably less than that expected from the laboratory studies. This discrepancy may be attributable to the fact that in the laboratory tunnel a full wave-field is unable to build up because of very limited fetch. Alternatively, it has been suggested (Quinn and Otto, 1971) that the square-law relationship found in the laboratory may not be a direct result of wind stress. Instead, the effect of wind may be to increase the rate of evaporation of water molecules, the resultant evaporative cooling of the surface water leading to convective mixing in the liquid near the interface. If convective mixing is important in promoting gas exchange, then in the natural environment it will depend on many processes other than wind velocity (e.g., air–water temperature difference, humidity). This could explain why laboratory results are often poor predictors of gas exchange rates at sea.

Various models have been developed that potentially allow $k_l$ to be calculated indirectly, such as the simple result used earlier in this paper (cf., Figure 4.8). More complicated models are the surface renewal model of Higbie (1935) and later developments of it (Danckwerts, 1951; Munnich, 1963). In these, the liquid near the interface is replaced intermittently by fluid from the bulk. Another class of models are those in which the surface water is described in terms of a regular system of eddies whose size is determined by the scale length of the turbulence in the underlying fluid (Fortescue and Pearson, 1967; Lamont and Scott, 1970). The main difficulty in applying all these models to environmental air–water interfaces, and especially the sea surface, is in specifying the necessary input parameters, e.g., the residence time of elements of fluid at the interface in the surface renewal models and the scale length and rate of turbulent dissipation of energy in the eddy models.

In order to gain a better understanding of $k_l$ for calculating gas exchange rates across the air–sea interface, the following experiments are of prime importance:

(a) Detailed, precise measurements of $k_l$ using the radon deficiency method in order to establish the relationship between the liquid-phase transfer velocity and meteorological parameters, such as wind speed. These measurements are probably best carried out from a weathership; the concentration of various gases in surface seawater and marine air should be measured at the same time.

(b) Careful laboratory wind-tunnel experiments should be carried out in order to identify the important factors controlling liquid-phase exchange mechanisms for gas transfer (e.g., evaporation–condensation, waves, spray, bubbles, wind, surface films).

Another, but possibly less important, problem connected with gas transfer across the air–sea interface is whether values for the Henry's law constant determined at relatively high partial pressures in the gas phase (often 1 atm of pure gas) are applicable to environmental situations. There is some evidence that for carbon monoxide (Meadows and Spedding, 1974) this is not the case. Laboratory determinations of $H$ at environmental concentrations should be carried out for gases whose air–sea exchange is important.

F.  PARTICLE DRY DEPOSITION TO THE OCEANS

Our knowledge of particle dry deposition to the oceans is severely restricted both experimentally and theoretically. Essentially the only relevant data available for deposition velocities as a function of particle size are shown in Figure 4.9, and these data were obtained from wind-tunnel studies (Sehmel and Sutter, 1974; Möller and Schumann, 1970). Under some conditions, wind-tunnel data are valuable, especially if the controlling resistance is in the deposition layer next to the surface; however, extrapolations from wind-tunnel data to estimates of the fluxes to the oceans must be suspect if, as was the case for the data of Figure 4.9, the state of the sea surface (including waves and spray) is not adequately duplicated. On the other hand, although some recent modeling studies are beginning to demonstrate understanding of particle dry deposition to rigid surfaces (Caporaloni *et al.*, 1975; Slinn, 1976b, 1976d) it appears that no one has yet successfully modeled the intricacies of particle dry deposition to the oceans. Thus, much future research is required both experimentally and theoretically; current research in both dry deposition and resuspension is well summarized in the recent conference proceedings edited by Engelmann and Sehmel (1976).

From Figure 4.8 and from the recent studies of Sehmel and Hodgson (1976), it does appear that the dominant resistance to particle dry deposition should occur in the last centimeter or so above the air–sea interface. For the case of dry deposition to rigid surfaces, Sehmel and Hodgson's calculations indicate that the resistance in this lowest layer is up to 3 orders of magnitude larger (for 0.1-$\mu$m particles) than the resistance in the layer from 1 cm to 10 m, almost regardless of atmospheric stability. A similar result can be deduced from Figure 4.8. Since the corresponding airborne concentrations of particles vary little above about 1 m, it would be convenient to report experimental deposition velocities as the ratio of the surface flux divided by the concentration at a 1-m or a 10-m reference height.

Some inferences about mass-average dry deposition velocities from radionuclide measurements are available but are not entirely satisfactory. Young and Silker's (1974) deposition velocities to the ocean of 0.7 to 2.2 cm sec$^{-1}$ for $^7$Be (presumably attached to maritime aerosol particles) may have been influenced by precipitation scavenging. Van der Hoven's (1968) review of the deposition velocities for various radionuclides to water surfaces are not specific either with respect to particle size or to details of the water surface characteristics. Results reported are as follows: $^{137}$Cs, 0.9 cm sec$^{-1}$; $^{103}$Ru, 2.3 cm sec$^{-1}$; $^{95}$Zr and $^{95}$Nb, 5.7 cm sec$^{-1}$, all values being substantially larger than the corresponding values given for soil, grass, or sticky paper.

Particle deposition to the ocean would be expected to increase with increasing wave height. This inference is made from Sehmel and Hodgson's (1976) correlation of wind-tunnel data on the increase in deposition to solid surfaces as the roughness length increases. Figure 4.10 shows the results of this correlation for $u_* = 30$ cm sec$^{-1}$ and a 1-m reference height. However, it must be re-emphasized that the roughness height $z_0$ is not related simply to the wave height (cf. earlier remarks or, e.g., see Businger, 1972; Gifford, 1976).

Any theoretical model of particle dry deposition to the oceans must

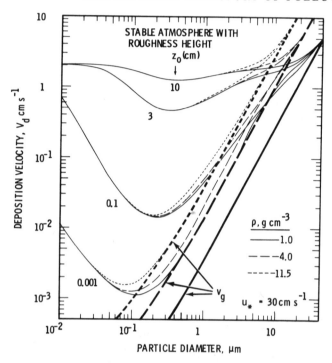

FIGURE 4.10 Correlation of experimental data by Sehmel and Hodgson (1976) for dry deposition of particles to various solid surfaces. The data encompassed roughness heights only up to about 0.1 cm, and, therefore, the extrapolation to greater roughness heights is tentative. The deposition velocity plotted is the flux divided by the extrapolated concentration at 1 m; $v_g$ is the gravitational settling speed for particles of indicated density; the case with $u_* = 30$ cm sec$^{-1}$ is shown ($\bar{u} \simeq 10$ m sec$^{-1}$); cases with other friction velocities can be found in the referenced publication.

include the effect of turbulence and surface roughness and probably should account for influences on particle motion from water-vapor condensation and evaporation, particle growth by water-vapor condensation, the induced motion of the sea surface, and particle capture by spray drops. To illustrate, in particular, the potential influence of water-vapor evaporation from the sea surface, consider an evaporating surface of water that remains at $z = 0$ in the $x$-$y$ plane; the restriction that the water plane remains at $z = 0$ bypasses difficulties associated with the motion of the water surface. Further, consider a fixed control volume, say of base area 1 cm$^2$ and height extending from $z = -1$ to

$z = +1$ cm, and let the water-vapor mass flux through this control volume be $\dot{m}_v''$ (positive for evaporation, negative for condensation). With this water-vapor flux there can, of course, be identified a drift velocity *of the water molecules*, $\dot{m}_v''/\rho_v$, where $\rho_v$ is the water vapor density. With this directed motion of water molecules, it can be expected that a drag force will act on the air molecules *and on any particles* within the control volume. The motion of the air molecules is not of much interest because after a short time (with characteristic velocity approximately the speed of sound) a density gradient of the air molecules will be established. The result will be compensating fluxes of air molecules (one flux dragged with the vapor, the opposite flux diffusing down the air's density gradient) with the total pressure (from the air plus the water vapor) a constant.

However, there will be a net flux of particles caused by the diffusion of the water vapor (Vittori and Prodi, 1967). This is known as Stefan flow or diffusiophoretic flux ("phoresis" = force). To determine the resulting velocity of the particles, it is noted first that the directed velocity $\dot{m}_v''/\rho_v$ is not the velocity of the whole fluid but only of the water molecules. After the density gradient in the air is established, the net velocity of the air molecules is zero. Consequently, the velocity of the fluid and, it is assumed, the velocity of any particles imbedded in the fluid, is the weighted mean of the two velocities. The resulting drift (or Stefan) velocity of the particles is

$$v_s = \frac{0 \cdot \rho_a + (\dot{m}_v''/\rho_v) \cdot \rho_v}{\rho_a + \rho_v} \simeq \frac{\dot{m}_v''}{\rho_a}, \qquad (4.74)$$

where the approximation to ignore $\rho_v$ is acceptable, since, for cases of interest, $\rho_v \ll \rho_a$. More detailed analyses give a slightly different result (Waldmann and Schmitt, 1966), but the differences from Eq. (4.74) are generally ignorable for the case of interest here. Numerically, it is seen that if $\dot{m}_v'' = 10$ g cm$^{-2}$ h$^{-1}$ (although, admittedly, 10 cm h$^{-1}$ is a rapid rate of evaporation) then from Eq. (4.74), $v_s \simeq 2$ cm sec$^{-1}$, which would essentially completely terminate the dry deposition of particles smaller than a few micrometers in size.

There are many other simultaneous processes that complicate the above simple picture. One factor is thermophoresis: the directed motion of a particle in the direction of any heat flux. On the one hand, it might be argued that if the evaporation flux were $\dot{m}_v''$ over a synoptic-scale region of the ocean then, after the temperature of the sea surface had dropped substantially, a significant portion of the latent heat

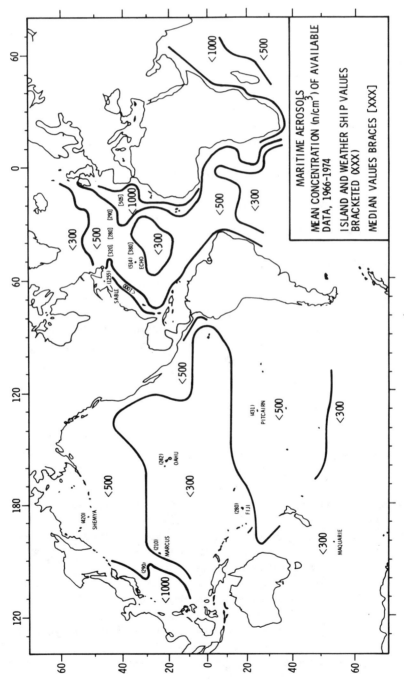

FIGURE 4.11 Geographic distribution of surface-level condensation (Aitken) nuclei measured over the oceans (Hogan, 1976a).

required for evaporation would be supplied by the atmosphere. The result would be a thermophoretic drift of the particles in the direction opposite to the diffusiophoretic drift; for the case of submicrometer particles and where all latent heat is supplied by the atmosphere, thermophoresis could overwhelm diffusiophoresis (Slinn and Hales, 1973). On the other hand, though, it is probably necessary to consider the microscale aspects of the evaporation process, wherein individual "eddies" sporadically are in contact with the surface, evaporation being intense only in some of these eddies, the corresponding heat required for evaporation is conducted from nearby regions of water, which, in turn, may receive heat at a less intense flux from a larger, neighboring region of the ocean and, ultimately, from solar radiation. To this picture must be added the turbulent impulses to the particles, which carry them across the viscous-dominated region of the atmosphere (Caporaloni *et al.,* 1975; Slinn, 1976b), and a consideration of where in the "eddies" these impulsive forces act on the particles compared with where diffusiophoresis and thermophoresis act.

To explore these concepts, measurements of particle deposition to wet surfaces should be accompanied by temperature and humidity profiles, roughness parameters, and heat and moisture fluxes, as well as the usual turbulence characterizations. Models such as these must be explored if an understanding of particle deposition to the sea surface is to be developed.

If the discussion above seems incomplete and unsatisfactory, it is because it reflects, to a considerable extent, our present knowledge. In spite of this, much information can be derived from measurements of the existing aerosol particles in the near-surface level, above the oceans. During the last few years, several thousand such observations have been made from islands, oceanographic ships, and merchant ships of opportunity. Figure 4.11 shows the resulting geographic distribution of mean, total aerosol number concentrations as measured by Aitken particle counters, which are sensitive to all particles greater than $\sim 0.001$ $\mu m$; such number concentrations are typically dominated by particles below 0.1 $\mu m$. It can be seen (Hogan, 1976a) that the concentration isopleths reflect continental sources and the effects of the dominant meteorological regimes. Continental tongues of higher concentration are found in some monsoonal areas, but, in general, the continental to maritime aerosol transition is quite rapid, as off the coast of North America. However, it should be noted that because these data are essentially number densities for very small particles, a significant fraction of the "removal" is probably attributable to interparticle coagulation rather than to deposition to the sea surface. There are indications that the rates of concentration changes for pollution aerosol particles of size near, or larger than, 0.1 $\mu$m is not nearly so rapid as the case illustrated in Figure 4.11 (Kojima and Sekikawa, 1974; Kojima *et al.*, 1974; Misaki *et al.*, 1975; Hogan 1976b). However, these conclusions, too, must be considered as tentative until complete evaluation is made of the contributions to the surface-level particle concentrations from air that has overridden the air adjacent to the sea surface.

From the presentations in this subsection, presumably it is clear that much is unknown, both from the experimental and theoretical viewpoints about particle dry deposition to the oceans. Tracer studies of particle deposition to the oceans or to lakes could be very productive; both Eulerian (or grid) and "Lagrangian" (or air-mass) studies should be considered. In lieu of ocean measurements, further research would be useful in wind tunnels, and new studies over large ponds could yield valuable data, especially if wave generation could be controlled and the resulting variations in deposition velocity studied. Much can also be learned from measurements of the existing maritime aerosol, especially if, for example, measurements were made of the evolutionary changes in the gas and particle phases as a polluted air mass emerges from a continent and passes over an ocean. More effort should also be devoted to model development and to investigating the validity of extrapolations from laboratory results to predictions about behavior in the ocean environment.

## VI.   CONCLUDING REMARKS

The purposes of this report have been to demonstrate some of the knowns and unknowns about the wet and dry fluxes of atmospheric trace constituents through the air–sea interface. Under the restrictions that have been discussed, the knowns can be summarized as follows. The dry flux for gases can be estimated as

$$D_d = k_t(\tilde{\chi}_b - H\tilde{C}_b), \qquad (4.75)$$

with the total transfer velocity given by

$$\frac{1}{k_t} = \frac{1}{k_g} + \frac{H}{\alpha^* k_l}, \qquad (4.76)$$

with $k_g$ and $k_l$ roughly as given in Figure 4.8. The dry deposition of particles can only be qualitatively estimated, using the deposition velocities shown in Figure 4.9. The wet flux can be estimated to be

$$W = rp_o\bar{\chi}_b, \qquad (4.77)$$

with the washout ratios as given in Tables 4.4 and 4.5. For scavenging of gases that form simple solutions in rainwater, $r = \alpha$; reactive pollutant gases are probably converted to their reaction products before being scavenged; and the washout ratios for particles are given in Eq. (4.29) using the removal rates in Eqs. (4.23) or (4.24).

From these relations and from the definitions of the average height from which pollution is removed by dry processes, as in Eq. (4.11), reservoir residence times of a pollutant can be estimated. For dry deposition from a layer of height $\bar{h}_d$ and with a deposition velocity $\bar{k}_t$, the residence time is

$$\bar{\tau}_d = \bar{h}_d/\bar{k}_t. \qquad (4.78)$$

The rain scavenging contribution to the residence time is as in Eq. (4.35); e.g.,

$$\bar{\tau}_w = \bar{R}_m/cp\bar{E}. \qquad (4.79)$$

To obtain these results, use has been made of simplifications that would be inappropriate for use near a specific pollutant source. If, in addition to wet and dry cleansing processes, other chemical or physical processes contribute to a pollutant's removal from the atmosphere, then the residence time is given by

$$\frac{1}{\bar{\tau}} = \frac{1}{\tau_d} + \frac{1}{\tau_w} + \frac{1}{\tau_o}; \qquad (4.80)$$

in other words, the separate removal paths act as resistances in parallel. As an example of the use of Eq. (4.80), consider an air parcel with aerosol particles having $\tau_d \simeq 10$ days and $\tau_w \simeq 10$ days, and where other processes (e.g., interparticle coagulation) proceed at a negligibly slow rate ($\tau_o \to \infty$), then Eq. (4.80) gives $\bar{\tau} \simeq 5$ days.

At a number of places in this chapter, the potential importance of other removal processes has appeared, but their details have been ignored. Examples include the oxidation of $SO_2$ to $SO_4^{2-}$, the coagulation of Aitken nuclei, and the attachment of radionuclides, pesticides, and other vapors to aerosol particles that are subsequently removed by precipitation. All these subject areas require further investigation. Recently, Junge (1975) discussed the attachment of some pesticides and PCB's to aerosols, but much remains to be done. In the case of low-molecular-weight halogenated hydrocarbons, it may be that the only important "removal" mechanism is by their (photo-) chemical destruction in the atmosphere. It is therefore to be emphasized that the focus in this chapter on wet and dry removal processes reflects more the authors' purposeful restriction on the scope of the paper rather than their opinions on the relative importance of various removal mechanisms.

## VII. RECOMMENDATIONS

In the course of this chapter, we have identified many unknowns and suggested areas for future research. Although we find it difficult to agree on a set of priorities, we do agree that the following research areas require special attention. The ordering of recommendations generally follows the order in which the topics appeared in the text.

1. Interreservoir exchange rates for large-scale reservoirs are not known to within a factor of 2 or 3 at best. More extensive measurements of the air concentrations and the deposition rates of various radionuclides, both natural and anthropogenic, could reduce this uncertainty.

2. Data on the removal of radionuclides from the atmosphere could be used to draw inferences about the behavior of aerosol particles in the transport and deposition cycle. However, before this can be done, we need more information on the character of the aerosols to which the radionuclides are attached, especially their size.

3. There is a general lack of storm precipitation statistics. The information required includes storm height, areal extent, frequency of occurrence, duration, precipitation amounts, and cloud-water removal efficiencies.

4. Much could be learned about scavenging efficiencies and modification of substances within clouds by performing budget studies for progressively more complex nonprecipitating and precipitating clouds and storms. Experiments should start with simple wave and cap clouds, progress to orographic and cumulus clouds, and culminate in a study of cumulonimbus and frontal storms.

5. Little is known about the dynamics of the air–sea interface. These

uncertainties are detrimental to our obtaining an understanding of dry deposition and resuspension of particles and gases. Needed are concerted scientific studies on the turbulence above and below the interface, wave mixing, the physics of wave breaking, gas and particle entrainment, particle resuspension, and thermally driven circulations above the thermocline.

6. For many gases, the dominant resistance to transfer occurs in the liquid phase and probably in the viscous sublayer. Wind-tunnel studies could help to establish the importance of evaporation and condensation, waves, spray, bubbles, wind speed, and surface films. The radon deficiency method holds most promise for determining the liquid-phase resistance in the ocean.

7. Particle dry deposition and resuspension are areas of major uncertainty in our knowledge of pollutant transfer to the oceans. To remove the uncertainties, studies on all fronts are recommended: controlled wind-tunnel studies, theoretical developments, semicontrolled deposition to ponds and lakes, and budget studies using both grid (Eulerian) and air-mass ("Lagrangian") frames over the open oceans.

## APPENDIX 4.A SYMBOLS

The following is a list of symbols used frequently in the text, their dimensions (L, length; T, time; M, mass), and, in case of multiple use of a single symbol, the equation numbers in which the symbol appears.

$a$ = particle radius, L
$A$ = hydrometeor cross-sectional area, $L^2$
$c$ = empirical constant $\simeq 0.5$, Eq. (4.23)
$c_o$ = peak phase velocity of gravity waves, $LT^{-1}$
$C$ = pollutant concentration in ocean, units $L^{-3}$
$C_d$ = drag coefficient, dimensionless
$\mathbf{d}$ = pollutant drift (or slip) velocity, $LT^{-1}$
$D$ = coefficient of Brownian or molecular diffusivity, $L^2T^{-1}$
$D_i$ = total rate of destruction within reservoir $i$, units $T^{-1}$
$D_d$ = magnitude of the dry deposition flux, units $L^{-2}T^{-1}$
$E_1$ = collection efficiency of particles by drops
$E_2$ = collection efficiency of particles by ice crystals
$\mathbf{F}$ = diffusive flux, units $L^{-2}T^{-1}$
$F_i$ = flow from reservoir $i$, units $T^{-1}$
$g$ = acceleration of gravity, $LT^{-2}$

$G$ = source (or gain) term, units $L^{-3}T^{-1}$

$h_d$ = height from which the pollutant is removed by dry deposition at rate $k_t$, L

$h_w$ = effective height from which the pollutant is removed by wet processes, L

$H$ = Henry's law constant, $\chi/\kappa$ or $\chi/C$, dimensionless

$I_i$ = inflow of pollutant to reservoir $i$, units $T^{-1}$

$I_{ij}$ = rate of inflow to reservoir $i$ from reservoir $j$, $T^{-1}$

$\tilde{K}$ = turbulent or eddy diffusivity, $L^2T^{-1}$

$K_z$ = $z$ component of turbulent diffusivity, $L^2T^{-1}$

$k$ = von Kármán's constant $\simeq 0.4$

$L$ = sink (or loss) term, units $L^{-3}T^{-1}$, Eq. (4.1)

$L$ = liquid or solid water content in cloud, $ML^{-3}$, Eq. (4.28)

$l$ = hydrometeor size parameter

$l_{ji}$ = rate of outflow from reservoir $i$ to reservoir $j$, $T^{-1}$

$\dot{m}_v''$ = water vapor mass flux, $ML^{-2}T^{-1}$

$N$ = hydrometeor number density function

$O_i$ = outflow of pollutant from reservoir $i$, units $T^{-1}$

$p$ = precipitation rate, $LT^{-1}$ or volume flux of precipitation, $L^3L^{-2}T^{-1}$

Pe = Peclet number = Re Sc, dimensionless

$P_i$ = total rate of production within reservoir $i$, units $T^{-1}$

$q$ = total amount of tracer still present in the atmosphere, units

$Q_i$ = quantity of pollutant in reservoir $i$, units

$r$ = washout ratio, $(\kappa/\chi)_o$, dimensionless, Eq. (4.29)

$r$ = resistance = $k^{-1}$, $TL^{-1}$, Eq. (4.42)

$R$ = drop radius, L

Re = Reynolds number, dimensionless

$S$ = Stokes number, dimensionless

$S_*$ = critical Stokes number, Figure 4.6

Sc = $\nu/D$, Schmidt number, dimensionless

$\bar{u}$ = mean wind speed, $LT^{-1}$

$u_*$ = friction velocity, $LT^{-1}$

$v$ = $\mu_w/\mu_a$ = ratio of dynamic viscosities, dimensionless, Figure 4.6

$\mathbf{v}$ = fluid velocity, $LT^{-1}$

$v_d$ = deposition velocity, $LT^{-1}$

$v_g$ = gravitational settling speed, $LT^{-1}$

$v_r$ = resuspension velocity, $LT^{-1}$

$v_s$ = magnitude of other slip velocities, e.g., Stefan velocity, $LT^{-1}$

$W$ = magnitude of the wet flux, units $L^{-2}T^{-1}$

$z$ = vertical coordinate, L

$z_o$ = roughness height, L

$\alpha$ = $H^{-1}$, solubility coefficient, dimensionless

$\alpha^*$ = effective solubility coefficient, accounting for irreversible reactions, dimensionless

$\beta$ = first-order reaction rate constant, $T^{-1}$

$\delta$ = thickness of atmospheric or oceanic layers

$\epsilon$ = fractional amount of pollutant removed by a storm

$\kappa$ = pollutant concentration in precipitation, units $L^{-3}$, Eq. (4.29)

$\kappa$ = $a/R$, interception parameter, dimensionless, Figure 4.6

$\lambda$ = decay rate for a particular radionuclide, $T^{-1}$

$\Lambda$ = average gas transfer rate to drops, $T^{-1}$

$\rho$ = mass density, $ML^{-3}$

$\sigma$ = rms wave height, L

$\tau$ = pollutant residence time, T, Eq. (4.8)

$\tau$ = particle stopping time, T, Figure 4.6

$\tau_d$ = residence time if only dry removal processes are acting, T

$\tau_i$ = lifetime, residence time, or turnover time of pollutant in reservoir $i$, T

$\tau_w$ = residence time if only wet removal processes are acting, T

$\nu$ = frequency with which pollutant encounters a precipitating storm, $T^{-1}$, Eq. (4.19)

$\nu$ = kinematic viscosity coefficient, $L^2T^{-1}$, Eq. (4.43)

$\psi$ = precipitation scavenging rate coefficient, $T^{-1}$

$\chi$ = pollutant concentration in air, units $L^{-3}$

$\chi_*$ = near-surface air concentration for dynamic equilibrium of deposition and resuspension, units $L^{-3}$

*Subscripts*

$a$ = air, atmosphere, aerodynamic

$A,B,...I$ = layer identification

$b$ = bulk

cw = cloud water

$d$ = dry

$g$ = gas phase

$i$ = interface, $i$th reservoir, initial

$l$ = liquid phase

$m$ = molecular, volume or mass mean

$n$ = normal
$o$ = ground level
$r$ = rain
$s$ = surface, snow
$ss$ = steady state
$t$ = total, terminal
$v$ = vapor
$w$ = water, wet
$z$ = vertical component

## Other Symbols

$\mathbf{r}, t$ = position vector, time
$\tilde{\xi}$ = average $\xi$ over microscale processes
$\bar{\xi}$ = average $\xi$ over large-scale meteorological processes
$\langle \xi \rangle$ = average $\xi$ over height of storm
$\xi'$ = the fluctuating component of $\xi$ which has zero microscale average
$\sum_j{}'$ = sum of $j$ with no sum on $i$

## REFERENCES

Adam, J. R., and R. G. Semonin (1970). Collection efficiency of raindrops for submicron particulates, see Engelmann and Slinn (1970).

Beadle, R. W., and R. G. Semonin (1976). *Precipitation Scavenging—1974*. Proceedings of a symposium held at Champagne, Ill., October 14–18, 1974; ERDA Symposium Series, to be available as CONF-741014 from NTIS, Springfield, Va.

Bowen, V. T., and T. T. Sugihara (1960). Strontium[90] in the "mixed-layer" of the Atlantic Ocean, *Nature 186*, 71–72.

Brimblecombe, P., and D. J. Spedding (1972). Rate of solution of gaseous sulphur dioxide at atmospheric concentrations, *Nature 236*, 225–229.

Brock, J. R. (1970). Attachment of trace substances on atmospheric aerosols, see Engelmann and Slinn (1970).

Brocks, K., and L. Krügermeyer (1972). The hydrodynamic roughness of the sea surface, *Studies Phys. Oceanog. 1*, 75–92.

Broecker, W. S. (1965). An application of natural radon to problems in ocean circulation, in *Symposium on Diffusion in Oceans and Fresh Waters*, T. Ichiye, ed., Columbia U. Press, New York, pp. 116–145.

Broecker, W. S. (1976). Personnal communication to P. S. Liss, U. of East Anglia, Norwich, U.K.

Broecker, W. S., and T. H. Peng (1974). Gas exchange rates between air and sea, *Tellus 26*, 21–35.

Browning, K. A., and G. B. Foote (1976). Air flow and hail growth in supercell storms and some implications for hail suppression, *Quart. J. R. Meteorol. Soc. 102*, 499–533.

Brutsaert, W. (1975). The roughness length for water vapor, sensible heat, and other scalars, *J. Atmos. Sci. 32*, 2028–2031.

Businger, J. A. (1972). Turbulent transfer in the atmospheric surface layer, *Workshop on Micrometeorology*, August 14–18, 1972, American Meteorol. Soc., Boston, Mass.

Calder, K. L. (1961). Atmospheric diffusion of particulate matter considered as a boundary value problem, *J. Appl. Meteorol. 12*, 413–416.

Cambray, R. S., E. M. R. Fisher, A. Parker, and D. H. Peirson (1973). *Radioactive Fallout in Air and Rain: Results to the Middle of 1973*, AERE Harwell Rep. R7540, HMSO, London.

Caporaloni, M., F. Tampieri, F. Trombetti, and O. Vittori (1975). Transfer of particles in nonisotropic air turbulence, *J. Atmos. Sci. 32*, 565–569.

Cawse, P. A. (1974). *A Survey of Atmospheric Trace Elements in the United Kingdom*, AERE Harwell Rep. R7669, HMSO, London.

Chamberlain, A. C. (1960). Aspects of the deposition of radioactive and other gases and particles, *Intern. Air Pollut. 3*, 63–88.

Corrsin, S. (1974). Limitations of gradient transport models in random walks and in turbulence, *Advan. Geophys. 18A*, 25–60.

Cox, C., and W. Munk (1955). Some problems in optical oceanography, *J. Marine Res. 14*, 63–78.

Crooks, R. N., R. G. D. Osmond, E. M. R. Fisher, M. J. Owers, and T. W. Evett (1960). AERE Harwell Rep. R3349, HMSO, London.

Dana, M. T. (1976). Battelle Northwest, Richland, Wash., personal communication to W. G. N. Slinn.

Dana, M. T., J. M. Hales, and M. A. Wolf (1975). Rain scavenging of $SO_2$ and sulfate from power plant plumes, *J. Geophys. Res. 80*, 4119–4129.

Danckwerts, P. V. (1951). Significance of liquid-film coefficients in gas adsorption, *Ind. Eng. Chem. 43*, 1460.

Dawson, D., and O. Trass (1972). Mass transfer at rough surfaces, *Intern. J. Heat Mass Transfer 15*, 1317–1336.

Deacon, E. L. (1973). Geostrophic drag coefficients, *Boundary Layer Meteorol. 5*, 321–340.

Dipprey, D. F., and R. H. Sabersky (1963). Heat and momentum transfer in smooth and rough tubes at various Prandtl numbers, *Intern. J. Heat Mass Transfer 6*, 329–353.

Dobson, F. W. (1971). Measurements of atmospheric pressure on wind-generated sea waves, *J. Fluid Mech. 48*, 91–127.

Downing, A. L., and G. A. Truesdale (1955). Some factors affecting the rate of solution of oxygen in water, *J. Appl. Chem. 5*, 570–581.

Dunckel, M., L. Hasse, L. Krügermeyer, D. Schriever, and J. Wucknitz (1974). Turbulent fluxes of momentum, heat and water vapor in the atmospheric surface layer at sea during ATEX, *Boundary Layer Meteorol. 7*, 363–372.

Dyer, A. J. (1974). A review of flux-profile relationships, *Boundary Layer Meteorol. 7*, 363–372.

Emerson, S. (1975). Chemically enhanced $CO_2$ gas exchange in a eutrophic lake: A general model, *Limnol. Oceanog. 20*, 743–753.

Engelmann, R. J. (1968). The calculation of precipitation scavenging, *Meteorology and Atomic Energy*, D. H. Slade, ed., USAEC Rep. TID-24190, available from NTIS, Springfield, Va., pp. 208–221.

Engelmann, R. J., and G. A. Sehmel (1976). *Atmosphere-Surface Exchange of Particulate and Gaseous Pollutants—1974*, Proceedings of a symposium held at Richland, Wash., Sept. 1974, ERDA Symposium Series 38, available as CONF-74-921 from NTIS, Springfield, Va.

Engelmann, R. J., and W. G. N. Slinn (1970). *Precipitation Scavenging—1970*, Proceedings of a symposium held at Richland, Wash., June 2–4, 1970, AEC Symp. Series 22, available as CONF-700601 from NTIS, Springfield, Va.

Engelmann, R. J., *et al.* (1966). *Washout Coefficients for Selected Gases and Particulates*, BNWL-SA-657, BNSA-155, BNWL-SA-798, Battelle-Northwest, Richland, Wash., available from NTIS, Springfield, Va.

Fortescue, G. E., and J. R. A. Pearson (1967). On gas absorption into a turbulent liquid, *Chem. Eng. Sci. 22*, 1163–1176.

Fukuda, K., and S. Tsunogai (1975). Pb-210 in precipitation in Japan and its implication for the transport of continental aerosols across the ocean, *Tellus 27*, 514–521.

Garrett, W. D. (1972). Impact of natural and man-made surface film on the properties of the air-sea interface, *The Changing Chemistry of the Oceans*, D. Dyrssen and D. Jagner, eds., Wiley-Interscience, New York, pp. 75–91.

Garratt, J. R., and B. B. Hicks (1973). Momentum, heat and water vapour transfer to and from natural and artificial surfaces, *Quart. J. R. Meteorol. Soc. 99*, 680–687.

Gatz, D. F. (1966). *Deposition of atmospheric particulate matter by convective storms*, Ph.D. Thesis and USAEC Rep. C00-1407-6, U. of Michigan.

Gatz, D. F. (1976a). Scavenging ratio measurements in METROMEX, see Beadle and Semonin (1976).

Gatz, D. F. (1976b). A review of chemical tracer experiments on precipitation systems, see Beadle and Semonin (1976).

Gedeonov, L. I., Z. G. Gritchenko, F. M. Flegontov, and M. I. Zhilkina (1970). Coefficient of radioactive-aerosol concentration in atmospheric precipitation, in *Atmospheric Scavenging of Radioisotopes*, B. Styra *et al.*, eds., Collection of papers delivered at a conference in Palanga, June 7–9, 1966. Proceedings available as TT 69-55099 from NTIS, Springfield Va.

Gifford, F. A. (1976). Turbulent diffusion typing schemes: a review, *Nuclear Safety 17*, 68–86.

Gillette, D. A., I. H. Blifford, and C. R. Fenster (1972). Measurements of aerosol size distributions and vertical fluxes of aerosols on land subject to wind erosion, *J. Appl. Meteorol. 11*, 977–987.

Hales, J. M. (1973). Fundamentals of the theory of gas scavenging by rain, *Atmos. Environ. 6*, 635–659.

Hales, J. M., and S. L. Sutter (1973). Solubility of sulfur dioxide in water at low concentrations, *Atmos. Environ. 7*, 997–1001.

Hasse, L. (1974). On the surface to geostrophic wind relationship at sea and the stability dependence of the resistance law, *Beitr. Phys. Atmos. 47*, 45–55.

Hasselmann, K., T. P. Barnett, E. Bouws, H. Carlson, *et al.* (1973). Measurements of wind-wave growth and swell delay during the Joint North-Sea Wave Project (JONSWAP), *Ergaenzungsh. zur Deutsch. Hydrogr. Z., Reihe A, Nr. 12*.

Heines, T. S., and L. K. Peters (1974). The effect of ground level absorption on the dispersion of pollutants in the atmosphere, *Atmos. Environ. 8*, 1143–1153.

Hicks, B. B., and P. S. Liss (1976). Transfer of $SO_2$ and other reactive gases across the air–sea interface, *Tellus 28*, 348–354.

Higbie, R. (1935). The rate of absorption of a pure gas into a still liquid during short periods of exposure, *Trans. Am. Inst. Chem. Eng. 35*, 365–373.

Hogan, A. W. (1976a). Physical properties of the atmospheric aerosol, Ph.D. Thesis, Dept. of Geophysics, Hokkaido U., Sapporo, Japan, April.

Hogan, A. W. (1976b). Aerosols of the trade wind region, *J. Appl. Meteorol. 15*, 611–619.

Hoover, T. E., and D. C. Berkshire (1969). Effects of hydration on carbon dioxide exchange across an air–water interface, *J. Geophys. Res. 74*, 456–464.

IITRE (1974). e.g., see E. O. Knutson and J. D. Stockham, Aerosol scavenging by snow-laboratory and field results, Beadle and Semonin (1976).

Junge, C. E. (1963). *Atmospheric Chemistry and Radioactivity*, Academic Press, New York.

Junge, C. E. (1964). The modification of aerosol size distributions in the atmosphere, *Final Tech. Rep.*, *Meteorol. Geophys. Inst.*, Johannes Gutenberg U., Contract DA-91-591-EVC-2979, available from NTIS, Springfield, Va.

Junge, C. E. (1975). Basic considerations about trace constituents in the atmosphere as related to the fate of global pollutants, presented at 169th ACS Natl. Mtg., Philadelphia, Pa., Apr. 7–11, Symposium on the Fate of Pollutants in the Air and Water Environment; published in *Advances in Environmental Science and Technology*, I. H. Suffet, ed. (1977), Vol. 8, Part I, Wiley-Interscience, New York, pp. 7–25.

Kanwisher, J. (1963). On the exchange of gases between the atmosphere and the sea, *Deep-Sea Res. 10*, 195–207.

Kerker, M., V. Hampl, D. D. Cooke, and E. Matijevic (1974). Scavenging of aerosol particles by a falling water droplet, *J. Atmos. Sci. 28*, 1211–1221.

Kitaigorodskii, S. A. (1970). *The Physics of Air–Sea Interaction*, translated from Russian and published by Israel Program for Scientific Translations Ltd., Jerusalem; available as TT72-50062 from NTIS, Springfield, Va.

Kojima, H., and T. Sekikawa (1974). Some characteristics of background aerosols over the Pacific Ocean, *J. Meteorol. Soc. Jpn. 52*, 499–504.

Kojima, H., T. Sekikawa, and F. Tanaka (1974). On the volatility of large particles in the urban and oceanic atmosphere, *J. Meteorol. Soc. Jpn. 52*, 90–92.

Lal, D., and Rama (1966). Characteristics of global tropospheric mixing based on man-made $C^{14}$, $H^3$, and $Sr^{90}$, *J. Geophys. Res. 71*, 2865–2874.

Lal, D., V. N. Nijampurkar, G. Rajagopalan, and B. L. K. Somayajulu (1976). Observations of fallout of natural radio-isotopes at tropical latitudes (in preparation).

Lamont, J. C., and D. S. Scott (1970). An eddy cell model of mass transfer into the surface of a turbulent liquid, *Am. Inst. Chem. Eng. J. 16*, 513–519.

Lassen, L., *et al.* (1960, 1961). For an English language review of their work, see Junge (1963).

Liss, P. S. (1973). Processes of gas exchange across an air–water interface, *Deep-Sea Res. 20*, 221–238.

Liss, P. S. (1975). Chemistry of the sea surface microlayer, in *Chemical Oceanography*, Vol. 2, J. P. Riley and G. Skirrow, eds., Academic Press, New York, pp. 193–243.

Liss, P. S. (1976). Effect of surface films on gas exchange across the air–sea interface, ICES Rapp. Proces Verb., in press.

Liss, P. S., and P. G. Slater (1974). Flux of gases across the air–sea interface, *Nature 247*, 181–184.

Martell, E. A., and H. E. Moore (1974). Tropospheric aerosol residence times: a critical review, *J. Rech. Atmos. III*, 903–910.

Marwitz, J. D. (1972). Precipitation efficiency of thunderstorms on the High Plains, *Preprints 3rd Conf. Weather Modification*, Rapid City, S.D., available from the Am. Meteorol. Soc., Boston, Mass., pp. 245–247.

Meadows, R. W., and D. J. Spedding (1974). The solubility of very low concentrations of carbon monoxide in aqueous solutions, *Tellus 26*, 143–150.

Misaki, M., M. Ikegami, and I. Kanazawa (1975). Deformation of the size distribution of aerosol particles dispersing from land to ocean, *J. Meteorol. Soc. Jpn. 53*, 111–120.

Möller, U., and G. Schumann (1970). Mechanisms of transport from the atmosphere to the earth's surface, *J. Geophys. Res. 75*, 3013–3019.

Munnich, K. O. (1963). Der Kreislauf des Radiokohlenstoffs in der Natur, *Naturwissenschaften 50*, 211–218.

Munnich, K. O. (1971). Atmosphere–ocean relationships in global environmental pollution, Special Environ. Rep. No. 2, *Meteorology as Related to the Human Environment*, WMO-No. 312, available from the Secretariat of the World Meteorological Organization, Geneva.

Owen, P. R., and W. R. Thomson (1963). Heat transfer across rough surfaces, *J. Fluid Mech. 15*, 321–334.

Peirson, D. H., P. A. Cawse, and R. S. Cambray (1974). Chemical uniformity of airborne particulate material, and a maritime effect, *Nature 251*, 675–679.

Postma, A. K. (1970). Effect of solubilities of gases on their scavenging by raindrops, see Engelmann and Slinn (1970).

Quinn, J. A., and N. C. Otto (1971). Carbon dioxide exchange at the air–sea interface: Flux augmentation by chemical reaction, *J. Geophys. Res. 76*, 1539–1549.

Roth, R. (1975). Der vertikale Transport van Luftbeimengungen in der Prandtl-Schicht und die Deposition-Velocity, *Meteorol. Rundsch. 28*, 65–71.

Schumann, G. (1975). The process of direct deposition of aerosols at the sea surface, preprint.

Schooley, A. H. (1969). Evaporation in the laboratory and at sea, *J. Marine Res. 27*, 335–338.

Sehmel, G. A. (1973). Particle eddy diffusivities and deposition velocities for isothermal flow over smooth surfaces, *J. Aerosol Sci. 4*, 125–133.

Sehmel, G. A., and W. H. Hodgson (1976). Particle dry deposition velocities, see Engelmann and Sehmel (1976).

Sehmel, G. A., and S. L. Sutter (1974). Particle deposition rates on a water surface as a function of particle diameter and air velocity, *J. Rech. Atmos. III*, 911–918.

Shemdin, O. H., and R. J. Lai (1970). Laboratory investigation of wave-induced motion above the air–sea interface, Dept. of Coastal and Oceanog. Eng., U. of Florida Tech. Rep. 6, 87 pp.

Shepherd, J. G. (1974). Measurements of the direct deposition of sulphur dioxide onto grass and water by the profile method, *Atmos. Environ. 8*, 69–74.

Sheriff, N., and P. Gumley (1966). Heat transfer and friction properties of surfaces with discrete roughnesses, *Intern. J. Heat Mass Transfer 9*, 1297–1319.

Slater, P. G. (1974). The exchange of gases across an air–water interface, Thesis, U. of East Anglia, 205 pp.

Slinn, W. G. N. (1976a). Formulation and a solution of the diffusion, deposition, resuspension problem, *Atmos. Environ. 10*, 763–768.

Slinn, W. G. N. (1976b). Some approximations for the wet and dry removal of particles and gases from the atmosphere, in *Proceedings of the First International Symposium on Acid Precipitation and the Forest Ecosystem*, L. S. Dochinger and T. A. Seliga, eds., USDA Forest Service Gen. Tech. Rep. NE-23, *J. Air Water and Soil Pollut. 7*, 513–543 (1977).

Slinn, W. G. N. (1976c). Precipitation scavenging: some problems, approximate solutions and suggestions for future research, see Beadle and Semonin (1976).

Slinn, W. G. N. (1976d). Dry deposition and resuspension of aerosol particles—a new look at some old problems, see Engelmann and Sehmel (1976).

Slinn, W. G. N., and J. M. Hales (1973). A re-evaluation of the role of thermophoresis as a mechanism of in- and below-cloud scavenging, *J. Atmos. Sci. 28*, 1465–1471.

Slinn, W. G. N., L. Hasse, B. B. Hicks, A. W. Hogan, D. Lai, P. S. Liss, K. O. Munnich, G. A. Sehmel, and O. Vittori (1977). Some aspects of the transfer of atmospheric trace constituents through the air–sea interface, *Atmos. Environ.*, in press.

Synder, R. L. (1974). A field study of wave-induced pressure fluctuations above surface gravity waves, *J. Marine Res. 32*, 497–531.

Sood, S. K., and M. R. Jackson (1972). Scavenging Study of Snow and Ice Crystals, Reps. IITRE C6105-9 (1969) and IITRI C6105-18 (1972), ITT Research Inst., Chicago, Ill.

Starr, J. R., and B. J. Mason (1966). The capture of airborne particles by water drops and simulated snow crystals, *Quart. J. R. Meterol. Soc. 92*, 490–497.

Stavitskaya, A. V. (1972). Capture of water-aerosol drops by flat obstacles in the form of star shaped crystals, *Izv. Akad. Nauk Atmos. Oceanic Phys. 8*, 768–774.

Styra, B., C. Garbaliauskas, and V. Lujanas (1970). *Atmospheric Scavenging of Radioisotopes*, available as TT69-55099 from NTIS, Springfield Va.

Van der Hoven, I. (1968). Deposition of particles and gases, *Meteorology and Atomic Energy—1968*, D. H. Slade, ed., available from NTIS, Springfield, Va.

Vittori, O., and V. Prodi (1967). Scavenging of atmospheric particles by ice crystals, *J. Atmos. Sci. 24*, 533–538.

Volchok, H. L. (1974). Is there excess $^{90}$Sr fallout in the oceans? U.S. Health and Safety Laboratory, Rept. No. HASL-286, Vol. I, pp. 82–88.

Volchok, H. L., V. T. Bowen, T. R. Folsom, W. S. Broecker, E. A. Schuert, and G. S. Bien (1971). Oceanic distribution of radionuclides from nuclear explosions, in *Radioactivity in the Marine Environment*, National Academy of Sciences, Washington, D.C.

Waldmann, L., and K. H. Schmitt (1966). Thermophoresis and diffusiophoresis of aerosols, *Aerosol Science*, C. N. Davies, ed., Academic Press, New York, pp. 137–162.

Whitney, R. P., and J. E. Vivian (1949). Absorption of sulfur dioxide in water, *Chem. Eng. Progr. 45*, 323–337.

Yaglom, A. M., and B. A. Kader (1974). Heat and mass transfer between a rough wall and turbulent fluid flow at high Reynolds and Peclet numbers, *J. Fluid Mech. 62*, 601–623.

Young, J. A., and W. B. Silker (1974). The determination of air–sea exchange and oceanic mixing rates using $^7$Be during BOMEX, *J. Geophys. Res. 79*, 4481–4493.

Zimin, A. G. (1964). Mechanisms of capture and precipitation of atmospheric contaminants by clouds and precipitation, in *Problems of Nuclear Meteorology*, I. L. Karol and S. G. Malskov, eds., translation of *Boprosy Yadernoi Meteorologii*, State Pub. House for Lit. in the Field of Atomic Sciences and Engineering, pp. 139–182, USAEC Rep. AEC-tr-6128-, available from NTIS, Springfield, Va.

# 5 Metals

## I. INTRODUCTION

The evaluation of the magnitude of man's input of metals to the atmosphere and the ocean is rendered difficult by the fact that the rates of mobilization of these metals by natural processes is not well known. In contrast to the situation with many other pollutants such as synthetic organic chemicals and radionuclides, materials that have no natural sources, the metals emitted by anthropogenic processes are not readily distinguishable from those emitted by natural process. Thus, it will be necessary to characterize the total global cycle of these metals before the impact of man's activities can be assessed.

The objectives then, in this chapter, are to summarize and evaluate the existing data on the concentration of metals in the troposphere, to identify possible sources, natural and anthropogenic, for these metals, and to estimate their fluxes to the oceans.

## II. CONCENTRATION OF METALS IN THE MARINE ATMOSPHERE

There is a dearth of good-quality data on the concentration of metals in the marine atmosphere. This dearth can be attributed primarily to the

---

Members of the Working Group on Metals were W. H. Zoller, *chairman*; R. Chesselet, R. Chester, R. A. Duce, E. D. Goldberg, J. Jedwab, C. C. Patterson, and D. H. Peirson.

difficulties involved in the collection of large-volume air samples free from contamination and the problem of accurately analyzing these metals in the samples at the very low concentrations that normally obtain. The analytical difficulties could be reduced by increasing the sample size. However, the probability of contamination by material advected from local sources increases with sampling duration.

In addition, the length of the sampling period might be dictated by the meteorological considerations. At the operational level, the local wind characteristics, such as the sea-breeze effect observed on many islands, can limit sampling times. More fundamentally, it may be desirable to relate sampling duration to a specific synoptic situation so that each sample is representative of an identifiable large-scale meteorological feature, i.e., a specific air-mass type or a front. (The meteorological criteria for determining sample duration are discussed in Chapter 3, Table 3.1.) In practice, the primary consideration in metals studied has been to obtain the maximum sample size consistent with the minimized possibility of contamination from local sources. Generally, the operational optimum is realized when samples are collected aboard ships or at coastal sites on islands over a maximum period of a day or two (under carefully selected and monitored wind conditions).

Reliable data on metals in the marine atmosphere have only been obtained during the last few years. These data are summarized in Table 5.1. We have excluded from this compilation all data that we believe do not conform to the sampling criteria discussed in the preceding paragraph (with one possible exception, which is noted in the table and discussed below).

The concentration of atmospheric trace substances are generally log normally distributed. For this reason, we report the geometric means in Table 5.1. However, if the flux of material between the atmosphere and the sea surface is proportional to its atmospheric concentration, then short periods of high atmospheric concentration may dominate the flux. Hence, the arithmetic mean concentration is most appropriate for flux calculations. In most cases, the arithmetic and geometric means are not very different; arithmetic means are presented only for the Bermuda and eastern tropical Atlantic data in Table 5.1. Concentration ranges are included to provide an indication of the variability at a given site.

For the purpose of comparison, we include in Table 5.1 the mean metal concentrations measured at an altitude of 600 m at sites 32 to 48 km downwind of major urban centers in the northeastern United States (Young *et al.*, 1975). These concentrations should be representative of

TABLE 5.1 Concentration of Metals in the Atmosphere (Units: $10^{-9}$ g m$^{-3}$ of air STP)[a,b]

| Urban Regions (1975)[1] | | | | Bermuda (1973)[2] | | | | | |
|---|---|---|---|---|---|---|---|---|---|
| Metal | Geo. Mn. | GSD | Range | Metal | Arith. Mn. | ASD | Geom. Mn. | GSD | Range |
| Na | 510 | 3 | 130–2300 | Na | 2000 | 1700 | 1600 | 3 | 200–8000 |
| Mg | 730 | 3 | 150–2030 | Mg | 300 | 200 | 200 | 2 | 30–900 |
| Al | 1600 | 2 | 340–3800 | Al | 500 | 600 | 140 | 6 | 3–3000 |
| Ca | 1200 | 3 | 410–6100 | Ca | 200 | 200 | 140 | 3 | 6–1100 |
| K | 400* | | | K | 200 | 200 | 120 | 3 | 17–1000 |
| Fe | 1700 | 3 | 380–4800 | Fe | 300 | 400 | 90 | 5 | 4–1900 |
| Pb | 170 | 3 | 48–1000 | Pb | 7 | 7 | 3 | 4 | 0.10–20 |
| Zn | 120 | 3 | 29–740 | Zn | 6 | 6 | 3 | 3 | 0.2–20 |
| Mn | 32 | 3 | 8–110 | Mn | 3 | 5 | 1.2 | 4 | 0.03–30 |
| V | 16 | 3 | 9–170 | V | 1.5† | 2 | 0.8 | 3 | 0.2–6 |
| Cu | 50* | | | Cu | 2 | 3 | 0.9 | 4 | <0.08–15 |
| Hg | 0.22 | 3 | 0.9–1.3 | Hg | | | | | |
| Cr | 14 | 4 | 2.6–153 | Cr | 0.5 | 0.7 | 0.3 | 3 | <0.04–3 |
| Ce | 3* | | | Ce | 0.6 | 0.8 | 0.2 | 5 | 0.005–3 |
| Cd | 3* | | | Cd | 0.4 | 0.4 | 0.2 | 4 | <0.01–1.6 |
| Se | 1.7 | 2.7 | 0.5–5.7 | Se | 0.19 | 0.17 | 0.13 | 3 | <0.02–0.6 |
| As | 16 | 2 | 7.5–50 | As | 0.12 | 0.12 | 0.07 | 3 | 0.012–0.5 |
| Co | 0.97 | 2.0 | 0.42–2.8 | Co | 0.08 | 0.11 | 0.03 | 5 | <0.005–0.5 |
| Sb | 3.0 | 2.9 | 0.81–12 | Sb | 0.03 | 0.03 | 0.014 | 5 | <0.001–0.3 |
| Sc | 0.39 | 2.3 | 0.11–1.3 | Sc | 0.06 | 0.09 | 0.02 | 5 | 0.002–0.4 |
| Th | 0.3 | | | Th | 0.05 | 0.08 | 0.02 | 5 | 0.002–0.2 |
| Ag | 0.6 | | | Ag | 0.009 | 0.02 | 0.003 | 3 | <0.002–0.08 |
| Eu | 0.056 | 2.8 | 0.016–0.21 | Eu | 0.009 | 0.012 | 0.003 | 5 | <0.0002–0.05 |

| Lerwick, Shetland Islands (1972)[5] | | | | Eastern Tropical Atlantic (1974)[6] | | | | | |
|---|---|---|---|---|---|---|---|---|---|
| Metal | Geo. Mn. | GSD | Range | Metal | Arith. Mn. | ASD | Geom. Mn. | GSD | Range |
| Na | 2000 | 2 | 600–5000 | Na | 2000 | 1200 | 1800 | 1.9 | 800–4000 |
| Mg | | | | Mg | 300 | 150 | 200 | 2 | 90–400 |
| Al | 60 | 1.9 | 20–150 | Al | 70 | 50 | 50 | 3 | 12–130 |
| Ca | | | | Ca | 150 | 50 | 150 | 1.4 | 80–200 |
| K | | | | K | 90 | 40 | 90 | 1.6 | 40–140 |
| Fe | 70 | 2 | 19–160 | Fe | 50 | 40 | 40 | 3 | 10–110 |
| Pb | 30 | 3 | <4–80 | Pb | 7 | 5 | 5 | 2 | 2–14 |
| Zn | 40 | 2 | 9–100 | Zn | 5 | 3 | 3 | 3 | 0.3–10 |
| Mn | 2 | 4 | 0.2–10 | Mn | 0.3 | 0.3 | 0.2 | 3 | <0.04–0.7 |
| V | 2 | 2 | 0.4–6 | V | 0.15 | 0.13 | 0.10 | 3 | <0.02–0.4 |
| Cu | <4 | | | Cu | 0.9 | 0.3 | 0.9 | 1.4 | 0.5–1.2 |
| Hg | <0.05 | | | Hg | 0.10 | 0.09 | 0.07 | 2 | <0.02–0.3 |
| Cr | 1.0 | 3 | 0.07–3 | Cr | 0.2 | 0.2 | 0.15 | 3 | 0.02–0.4 |
| Ce | 0.08 | 2 | 0.02–0.18 | Ce | 0.17 | 0.12 | 0.13 | 2 | 0.04–0.3 |
| Cd | <9 | | | Cd | | | | | |
| Se | 0.5 | 1.9 | <0.1–0.9 | Se | 0.3 | 0.11 | 0.3 | 1.4 | 0.2–0.5 |
| As | 1.3 | 4 | 0.14–8 | As | | | | | |
| Co | 0.07 | 2 | 0.03–0.18 | Co | 0.04 | 0.02 | 0.03 | 2 | 0.01–0.07 |
| Sb | 0.4 | 3 | 0.09–1.7 | Sb | 0.16 | 0.15 | 0.11 | 3 | 0.03–0.5 |
| Sc | 0.015 | 1.8 | 0.005–0.03 | Sc | 0.019 | 0.014 | 0.014 | 3 | 0.003–0.04 |
| Th | <0.04 | | | Th | 0.016 | 0.013 | 0.011 | 3 | 0.004–0.04 |
| Ag | | | | Ag | 0.05 | 0.10 | 0.014 | 5 | <0.007–0.3 |
| Eu | | | | Eu | | | | | |

| Northern Norway (1971–1972)[3] | | | | Gulf of Guinea (1970)[4] | | | |
|---|---|---|---|---|---|---|---|
| Metal | Geom. Mn. | GSD | Range | Metal | Geom. Mn. | GSD | Range |
| Na | 300 | 2 | 60–1500 | Na | | | |
| Mg | 60 | 2 | 12–180 | Mg | | | |
| Al | 40 | 2 | 6–130 | Al | | | |
| Ca | 40 | 1.6 | 15–90 | Ca | 1700 | 1.6 | 800–3000 |
| K | 40 | 1.6 | 15–80 | K | | | |
| Fe | 50 | 1.7 | 11–100 | Fe | 120 | 1.3 | 80–170 |
| Pb | 4 | 2 | 0.6–20 | Pb | | | |
| Zn | 7 | 1.9 | 1.7–30 | Zn | | | |
| Mn | 2 | 1.8 | 0.6–6 | Mn | | | |
| V | 1.2 | 3 | 0.19–5 | V | | | |
| Cu | 2 | 1.8 | 0.5–5 | Cu | | | |
| Hg | | | | Hg | 0.2 | 2 | 0.04–0.4 |
| Cr | 0.5 | 2 | 0.18–2 | Cr | 0.6 | 3 | 0.11–3 |
| Ce | 0.05 | 2 | 0.016–0.3 | Ce | 0.17 | 3 | 0.02–0.6 |
| Cd | 0.11 | 2 | 0.02–0.8 | Cd | | | |
| Se | 0.2 | 1.7 | 0.10–0.5 | Se | | | |
| As | 1.2 | 3 | 0.19–8 | As | | | |
| Co | | | | Co | 0.13 | 1.8 | 0.05–0.3 |
| Sb | 0.3 | 2 | 0.05–1.2 | Sb | 0.4 | 3 | 0.10–3 |
| Sc | 0.006 | 1.9 | 0.0016–0.02 | Sc | 0.02 | 1.4 | 0.017–0.04 |
| Th | 0.008 | 2 | 0.002–0.06 | Th | 0.02 | 1.6 | 0.009–0.04 |
| Ag | | | | Ag | | | |
| Eu | | | | Eu | 0.006 | 1.4 | 0.004–0.010 |

| Oahu, Hawaii (1969–1970)[7] | | | | South Pole (1974)[8] | | | |
|---|---|---|---|---|---|---|---|
| Metal | Geom. Mn. | GSD | Range | Metal | Geom. Mn. | GSD | Range |
| Na | 3000 | 1.7 | 900–14000 | Na | 3 | 1.5 | 1.7–4 |
| Mg | 400 | 1.7 | 0.10–1.8 | Mg | 0.7 | 1.8 | 0.3–1.4 |
| Al | 4 | 3 | 0.5–50 | Al | 0.6 | 1.9 | 0.2–1.4 |
| Ca | 140 | 1.6 | 0.06–0.7 | Ca | 0.5 | 1.5 | 0.3–0.8 |
| K | 100 | 1.7 | 0.03–0.3 | K | 0.7 | 1.2 | 0.5–0.9 |
| Fe | 9 | 2 | 1.0–50 | Fe | 0.5 | 1.6 | 0.3–1.0 |
| Pb | 2 | 2 | 0.3–13 | Pb | 0.03‡ | 1.4 | 0.02–0.04 |
| Zn | | | | Zn | 0.03 | 1.7 | 0.016–0.07 |
| Mn | 0.16 | 2 | 0.02–0.6 | Mn | 0.012 | 1.6 | 0.006–0.02 |
| V | 0.14 | 1.9 | 0.04–0.7 | V | 0.0013 | 1.5 | 0.0006–0.0016 |
| Cu | 1.4 | 2 | 0.2–12 | Cu | 0.04 | 1.6 | 0.02–0.05 |
| Hg | | | | Hg | | | |
| Cr | | | | Cr | 0.0043‡ | 1.7 | 0.003–0.010 |
| Ce | | | | Ce | 0.0019‡ | 1.9 | 0.0008–0.005 |
| Cd | | | | Cd | | | |
| Se | | | | Se | 0.006‡ | 1.2 | 0.005–0.007 |
| As | | | | As | 0.007‡ | 1.1 | 0.0067–0.0073 |
| Co | | | | Co | 0.0006 | 1.2 | 0.0005–0.0006 |
| Sb | | | | Sb | 0.0008 | 1.5 | 0.0005–0.0014 |
| Sc | | | | Sc | 0.00012 | 1.9 | 0.00006–0.0003 |
| Th | | | | Th | 0.00012 | 1.6 | 0.00007–0.0002 |
| Ag | | | | Ag | 0.002 | 3 | 0.0007–0.009 |
| Eu | | | | Eu | 0.000019 | 1.4 | 0.000013–0.00088 |

*See overleaf for footnotes.*

major coastal urban areas, which may well be major sources of the anthropogenic component of the metal flux to the oceans. Also presented are data from samples collected at the South Pole, this site being one of the most remote on the earth's surface. In general, the mean values for most elements in the North Atlantic samples are much less than the urban values but much higher than those from the Antarctic.

There are no data for the open ocean in the southern hemisphere that qualify for inclusion in Table 5.1. This is a serious deficiency. Data from the southern hemisphere would enable us to better assess anthropogenic impacts, because only 10 percent of the particulate pollutant sources are located in that hemisphere (Robinson and Robbins, 1971). The comparison of northern and southern hemisphere concentrations has been a valuable method for distinguishing between natural and anthropogenic sources for other trace substances such as carbon monoxide (Seiler, 1974; Newell *et al.*, 1974).

## III. SOURCES OF METALS IN THE MARINE ATMOSPHERE

### A.   REFERENCE ELEMENTS AND ENRICHMENT FACTORS

There are many possible natural sources for the metals present in the atmosphere; consequently, the anthropogenic contribution to the total

---

Footnotes to Table 5.1.

[a] Sample collection notes:

[1] Young *et al.* (1975); *Zoller, personal communication.

[2] Duce *et al.* (1976a; 1976b): Samples collected for one- to two-day periods from a coastal tower 40 m above sea level. Sample collection controlled automatically by wind speed and direction. The vanadium data† are from ship-collected samples near Bermuda (Duce and Hoffman, 1976). The number of samples analyzed was 29–65.

[3] Rahn (1976): Samples collected for one- to two-week periods at Skoganvarre near ground level. Collection location about 75 km from the open sea. No control of sampling. The number of samples analyzed was 21.

[4] Crozat *et al.* (1973): Samples collected from a ship for one-day periods on a 5-m-high bow mast, about 8 m above water level. Samples collected only when the ship was heading into the wind. The number of samples analyzed was 9.

[5] Peirson *et al.* (1974): Samples collected for one-month periods 0.5 km from the sea, near ground level but 75 m above sea level. No control of sampling. The number of samples analyzed was 11.

[6] Chesselet *et al.* (1975): Samples collected from a ship for one-day periods on a bow tower extending 6 m forward of ship and approximately 8 m above water level. Sample collection controlled automatically by wind direction. The number of samples analyzed was 8.

[7] Hoffman *et al.* (1972); Hoffman and Duce (1972): Samples collected for one-day periods from a windward coastal tower 20 m above sea level. Samples collected only during extended continuous periods of onshore tradewinds. The number of samples analyzed was 56–119.

[8] Maenhaut *et al.* (1977); ‡Zoller *et al.* (1974): Samples collected for approximately one-week sampling periods 6 m above ice level at an elevation of 2800 m above sea level. Collection controlled automatically by wind speed and direction and condensation nuclei counts. Data presented are generally from samples collected on Whatman 541 filters. Higher concentrations were observed for Pb, Se, and Sb in samples collected with 0.4-$\mu$m Nuclepore filters, suggesting that these elements are primarily present on very small particles. The number of samples analyzed was 4–10.

[b] In column head: GSD, geometric standard deviation; ASD, arithmetic standard deviation.

burden is not easily assessable. One approach to resolving the inputs from the various sources is to attempt to identify reference elements that are characteristic, or indicative, of specific natural sources. Ideally, a reference element must be amenable to relatively simple and accurate analysis, and it must be present in high concentration in aerosols produced by the source for which it serves as a reference and in low concentration in aerosols from other sources.

There is general agreement that the major portion of the aerosol mass over the oceans is derived from two sources: the sea itself, as spray from the bursting of bubbles, and the earth's crust, as mobilized soil material. Although a number of major constituents of seawater have been used as reference elements for that source, Na is generally preferred. Commonly used as reference elements for crustal weathering products are Al, Fe, Si, and Sc.

Of these, Si is probably best suited for this purpose. However, Al is used often because the analytical procedure is simpler and more accurate than that for Si.

For any element in an aerosol sample, we can define an enrichment factor relative to a reference element in a specific source. For seawater, using Na as the reference element, the enrichment factor, $EF_{sea}$, for an element, $X$, is

$$EF_{sea} = (X/Na)_{air} / (X/Na)_{sea}, \qquad (5.1)$$

where $(X/Na)_{air}$ and $(X/Na)_{sea}$ refer, respectively, to the ratio of the concentration of metal $X$ to that of Na in the atmosphere and in bulk seawater. Likewise, for crustal weathering products,

$$EF_{crust} = (X/Al)_{air} / (X/Al)_{crust}. \qquad (5.2)$$

If the concentration of any element in an air sample yields an $EF_{sea}$ or an $EF_{crust}$ value that is close to unity, then one can assume that the most probable source for that element in the aerosol was, respectively, seawater or crustal material. Some caution must be exercised in interpreting enrichment factors. With respect to $EF_{sea}$ values, the principal assumption is that the composition, relative to Na, of atmospheric sea-salt particles produced by bursting bubbles at the air–sea interface is identical to the composition of bulk seawater. However, it has been shown that this is not the case for many substances, such as iodine (Seto and Duce, 1972), phosphate (MacIntyre and Winchester, 1969), and total organic carbon (Barger and Garrett, 1970; Hoffman and Duce, 1976). These substances are significantly enriched in atmospheric sea-salt particles relative to bulk seawater. This fractionation is due, in part, to the association of these substances with surface-active organic compounds, which are scavenged by rising bubbles and transported to the water–atmosphere interface. Thus, the oceanic surface microlayer, which serves as the source for some of the material in the atmospheric sea-salt particles, becomes enriched with these materials relative to subsurface seawater. Many metals are known to be concentrated in the top several hundred micrometers of the air–sea interface (Piotrowicz et al., 1972; Szekielda et al., 1972; Barker and Zeitlin, 1972). It is expected that these metals will also be enriched in any sea-salt particles that incorporate material from the surface microlayer. The magnitude of the enrichment for these metals and its geochemical significance remain subjects of debate (Chesselet et al., 1976; Peirson et al., 1974; Van Grieken et al., 1974). However, Duce et al. (1976b) have presented strong evidence that Fe, Zn, and Cu are enriched by a factor of several hundred on atmospheric sea-salt particles produced by bubbles bursting in coastal waters.

Crustal enrichment factors do not enable us to distinguish between that crustal material which is injected into the atmosphere as a consequence of natural processes and that which is mobilized as a result of man's activities. The latter include increased exposure and breaking of soil surfaces for agricultural usage and in land-clearing operations; the production of crushed stone, sand, and gravel; and the suspension of soil particles by vehicular traffic. The interpretation of EF values is further complicated by the fact that many anthropogenic activities yield products (for example, coal and fly ash) in which many elements have relative concentrations that are similar to those of average crustal material. Moreover, one would not necessarily expect naturally derived soil aerosols to have a composition identical to that of average crustal material. First, the composition of rocks and soils varies from region to region. Second, the composition of the mobilized fraction of the soil may be different from that of the bulk soil material. Thus, $EF_{crust}$ ratios greater or less than unity do not necessarily preclude a crustal source.

Despite these limitations, the determination of $EF_{sea}$ and $EF_{crust}$ values can yield useful information on possible ocean and crustal sources for many metals. Enrichment factor values much greater than unity—a factor of 10 or more—alert us to metals that may be derived from other sources. Metals in the latter group, those that are apparently nonmarine and noncrustal in origin, may have major anthropogenic sources. It is these metals that ultimately may be of greatest concern.

## B. THE OCEAN

With the exception of the alkali and alkaline earth elements, the $EF_{sea}$ values for all the metals listed in Table 5.1 are very large, with values ranging from hundreds to tens of thousands. In contrast, the $EF_{sea}$ values for Mg, K, and Ca are generally in the range of $1 \pm 0.1$ in aerosols collected on the coast of Hawaii and from ships over the North Atlantic (Hoffman and Duce, 1972; Hoffman *et al.*, 1974). Slightly higher $EF_{sea}$ values for these metals have been observed (Chesselet *et al.*, 1972; Buat-Menard *et al.*, 1974). These higher values could be attributable to the presence of significant quantities of crustal aerosols in the atmosphere in these regions. However, there is some evidence that $EF_{sea}$ values for K may be higher in atmospheric sea-salt particles generated in regions of upwelling water (Buat-Menard *et al.*, 1974), possibly as a consequence of the increased biological productivity.

## C. CRUSTAL WEATHERING

A compilation of representative $EF_{crust}$ values is presented in Table 5.2. These data suggest that weathered crustal material is the primary source of Sc, Fe, Mn, Ce, Eu, and Co (i.e., their $EF_{crust}$ values are considerably less than 10). However, a number of elements are highly enriched relative to both the crust and seawater; these are Zn, Cu, Cd, Hg, Sb, Ag, As, Pb, and Se. Interest centers on these elements because of the possibility that they might have a significant anthropogenic component. Zoller *et al.* (1974) pointed out that most of these highly enriched elements, and their oxide or halide compounds, have a relatively low boiling point.

If a significant fraction of the earth-atmosphere flux of these metals involves a vapor phase, then we would expect their concentration to be relatively greater in submicrometer particles. The aerosol surface area is greatest in the submicrometer-size range, and, consequently, the effects of sorption, condensation, and surface reactions should be most apparent on these particles. This size dependency has been most clearly demonstrated in studies of aerosol composition and size distribution in urban areas. Generally, it is found that the total aerosol mass distribution is essentially bimodal (except in the immediate vicinity of major sources) with a saddle point at approximately 2-$\mu$m diameter (Whitby *et al.*, 1975; Sverdrup *et al.*, 1975). Particles larger than a few micrometers in diameter are generated by natural and anthropogenic mechanical (comminution) processes, while most of the particles below

TABLE 5.2  Mean Crustal Enrichment Factors for Metals in Atmospheric Particles

| Element | Urban Regions[a] | Lerwick Shetland Is. (1972)[b] | Bermuda (1973)[c] | Eastern Tropical Atlantic (1974)[d] | North Coast of Norway (1971-1972)[e] | Gulf of Guinea (1970)[f] | Oahu, Hawaii (1969-1970)[g] | South Pole (1974)[h] |
|---|---|---|---|---|---|---|---|---|
| Sc | 0.8 | 0.9 | 1.1 | 1.2 | 0.6 | 0.8 | — | 0.8 |
| Al | 1.0 | 1.0 | 1.0 | 1.0 | 1.0 | 1.0 | 1.0 | 1.0 |
| Fe | 1.2 | 1.7 | 1.0 | 1.1 | 1.9 | 1.7 | 2.6 | 1.2 |
| Th | 0.7 | 3.8 | 2.1 | 2.4 | 1.8 | — | — | 1.6 |
| Mn | 1.4 | 4.6 | 0.7 | 0.5 | 5.0 | — | 2.6 | 1.4 |
| Ce | 1.1 | 1.7 | 3.6 | 4.3 | 2.3 | 2.6 | — | 4.4 |
| Eu | 3.8 | — | 2.5 | — | — | — | — | 1.9 |
| Co | 1.8 | 2.8 | 1.8 | 2.4 | 6.3 | 4.0 | — | 2.6 |
| Cr | 14 | 15 | 1.7 | 3.2 | 12 | 6.8 | — | 6.9 |
| V | 17 | 24 | 17 | 1.8 | 24 | — | 15 | 1.1 |
| Zn | 200 | 790 | 26 | 480 | 240 | — | — | 50 |
| Cu | 21 | 77 | 12 | 22 | 79 | — | — | 84 |
| Pb | 1,800 | 3,300 | 180 | 720 | 800 | — | 450 | 240 |
| As | 89 | 1,500 | 50 | — | 1,900 | — | 2,900 | 510 |
| Ag | 200 | — | 52 | 1,400 | — | — | — | — |
| Hg | 570 | <700 | ≥65 | 1,800 | — | 1,800 | — | 2,600 |
| Sb | 910 | 3,600 | 180 | 3,700 | 3,600 | 2,400 | — | — |
| Cd | 340 | — | 570 | — | 1,200 | — | — | 410 |
| Se | 3,600 | 12,000 | 2,600 | 17,000 | 10,000 | — | — | 16,000 |

[a] Young et al. (1975); Zoller (personal communication).
[b] Peirson et al. (1974).
[c] Duce et al. (1976b), except for V, which is from Duce and Hoffman (1976).
[d] Chesselet et al. (1975).
[e] Rahn (1976).
[f] Crozat et al. (1973).
[g] Hoffman et al. (1972).
[h] Maenhaut et al. (1977), except for Ce and Cr, which are from Zoller et al. (1974).

132

a few micrometers in diameter are produced either directly in combustion processes or indirectly from the condensation of chemical or photochemical reaction products. In urban areas, metals in the small-particle mode are generally characterized by large $EF_{crust}$ values; examples are Cu, As, Hg, Zn, Sb, and Pb (Paciga and Jervis, 1976). In contrast, metals concentrated in the large-particle-size fraction (Fe, Sc, Co, and Mn) have low $EF_{crust}$ values (Paciga and Jervis, 1976) and are assumed to be soil-derived.

There are few data on the size distribution of metals in marine aerosols. However, one major study (Duce *et al.*, 1976a) has shown that, in aerosols collected at Bermuda, the metals having high $EF_{crust}$ values were concentrated in the submicrometer-size range, whereas the oceanic and crustal-source metals were concentrated in the larger particles. Also, supporting evidence comes from an aerosol size study made in coastal cities (Johansson *et al.*, 1974); these data show that the size dependence of the metal concentrations in air masses with an oceanic trajectory is essentially the same as that in air masses with a continental trajectory.

It would appear, then, that the metals having high $EF_{crust}$ values are derived from the continents and are products of vaporization–condensation/sorption processes. These volatilization processes could be anthropogenic; however, there are a number of natural processes, some of which will be discussed below, that possibly could produce the same results.

## D. POLLUTION

There are several major anthropogenic sources of atmospheric heavy metals: the combustion of fossil fuels (coal, gasoline, lignite, and oils), the incineration of waste, and the emissions from cement plants and other industrial sources. The fluxes of some heavy metals to the atmosphere from these sources can approach, or even exceed, natural fluxes to the marine environment (Study Panel on Assessing Potential Ocean Pollutants, 1975). For some individual heavy metals, single dominant sources often can be identified in certain regions. Examples include lead from the combustion of lead alkyls in gasoline (Murozumi *et al.*, 1969), vanadium from residual oil combustion (Zoller *et al.*, 1973), arsenic from smelters (Crecelius *et al.*, 1974), and cadmium from incinerators (Greenberg *et al.*, 1977). Many of these metals are released to the atmosphere from high-temperature combustion processes; these metals are primarily associated with submicrometer particulate material in urban areas as discussed above.

There are two categories of discrete particles found in the air and the sea that are exclusively man-made and that can easily and unambiguously be identified by electron microscope and microprobe techniques. These are automobile exhaust particles and inorganic pigments. These materials are stable in air and water and, because of their small size and long residence times, are widely distributed. Because the annual production figures for these substances are a matter of record, they can be used as tracers for the study of the dispersion of particulate anthropogenic materials. As yet, exhaust and pigment particles have not been systematically studied, and there are few data on their fluxes, their concentrations at various levels in the atmosphere and oceans, or their geographical distributions.

### E.   OTHER SOURCES

Although the concentrations of metals in the atmosphere in remote regions can vary by several orders of magnitude, the $EF_{crust}$ values generally do not. The similarities between the $EF_{crust}$ values at several of the widely distributed sites given in Table 5.2 have been noted and discussed by Duce et al. (1975); they suggest that the geographical uniformity of $EF_{crust}$ values may indicate that natural sources or processes could be responsible for some of the observed enrichments. Possibilities include the following.

### 1.   Low-Temperature Volatility and Crustal Degassing

Low-temperature vaporization of some metals or their compounds from crustal rocks has been postulated to take place (Goldberg, 1976). Degassing of the earth's crust may be the major source of atmospheric mercury (Weiss et al., 1971).

### 2. Biological Mobilization

Life processes can mobilize heavy metals to the atmosphere. For example, the methylation of arsenic and selenium by land plants and the subsequent release of the methyl arsenides and methyl selenides has long been known. Wood (1974) points out that methylated forms of As, Hg, S, and Se can be produced by microorganisms in the marine environment. More recently, evidence has been obtained that suggests that growing plants release zinc compounds to the atmosphere

(Beauford *et al.*, 1977). The importance of biological mobilization is yet to be established on a global basis.

## 3. Volcanic Activity

Heavy metals are introduced to the atmosphere intermittently through volcanic activity. Intense volcanic eruptions can directly inject particulate and gaseous materials into the stratosphere as well as the troposphere; as a consequence, these materials can be rapidly mixed throughout the atmosphere. Estimates of the mass yield of particles from volcanoes vary widely, but recent measurements suggest that yields could be quite large (Hobbs *et al.*, 1977). The metal output of volcanoes via the high-temperature volatilization mechanisms is unknown. However, the metals with high $EF_{crust}$ values in Table 5.2 also have very high $EF_{crust}$ values in particles collected in the fumaroles and vents of active volcanoes (Cadle *et al.*, 1973; Mroz and Zoller, 1975; Duce *et al.*, 1976b).

### F. METAL FLUXES TO THE ATMOSPHERE

One approach to understanding the origins of metals in remote areas is to estimate the global fluxes of each metal to the atmosphere from known sources to ascertain if any one source could be dominant. To this end, we have assembled calculations for the total fluxes of metals from crustal material, sea salt, and fossil fuels to the atmosphere (Table 5.3). It should be noted, as mentioned previously, that there may be other natural and anthropogenic sources for these aerosol metals. However, there is currently no information on which to base a calculation of the global fluxes of metals from these other sources.

The total mass flux of crustal material to the atmosphere has been estimated to be $2.5 \times 10^{14}$ g yr$^{-1}$ (Goldberg, 1971); the average crustal abundances of Taylor (1964) were used to convert this total crustal mass flux to elemental fluxes. The flux of sea-salt particles to the atmosphere is estimated to be $1 \times 10^{15}$ g yr$^{-1}$ (Eriksson, 1959). The sea-salt flux is converted to a metal flux using the average seawater concentrations reported by Riley (1975) and by Chester and Stoner (1974). This calculation assumes that no element fractionation occurs during sea-salt particle production. The flux of heavy metals from the combustion of coal, lignite, oil, and natural gas is taken from Bertine and Goldberg (1971). The flux of submicrometer lead to the atmosphere from anthropogenic processes (primarily the combustion of leaded fuels) is taken from Patterson *et al.* (1976).

TABLE 5.3    Global Flux of Metals to the Atmosphere Based on Total Crustal Material Flux of $2.5 \times 10^{14}$ g/yr[a] and Total Sea-Salt Flux of $1 \times 10^{15}$ g/yr[b]

| Element | Crustal Material[c] $10^9$ g/yr | Bulk Sea Salt[d] $10^9$ g/yr | Fossil-Fuel Combustion Products[e] $10^9$ g/yr |
|---|---|---|---|
| Al | 20,000 | 0.15 | 1400 |
| Fe | 14,000 | 0.5 | 1400 |
| Na | 6,000 | $3 \times 10^5$ | 300 |
| Mn | 200 | 0.005 | 7 |
| Sc | 6 | 0.000015 | 0.7 |
| Cu | 14 | 0.04 | 2 |
| V | 30 | 0.05 | 12 |
| Se | 0.013 | 0.003 | 0.5 |
| Pb | 3 | 0.0008 | 150[f] |
| Cd | 0.05 | 0.0008 | — |
| As | 0.5 | 0.05 | 0.7 |
| Zn | 18 | 0.08 | 0.5 |
| Sb | 0.05 | 0.007 | — |
| Hg | 0.02 | 0.0005 | 1.6 |

[a] Goldberg (1971).
[b] Eriksson (1959).
[c] Using crustal abundances of Taylor (1964).
[d] Using seawater concentrations of Riley (1975) and Chester and Stoner (1974).
[e] From estimates of Bertine and Goldberg (1971).
[f] Estimate of Patterson et al. (1976).

It is clear, from Table 5.3, that the crustal sources dominate the sea salt and fossil-fuel sources for most metals. The notable exceptions are Na from the oceans and V, Se, Hg, and Pb from the combustion of fossil fuels and leaded gasoline.

## IV.    FLUXES OF METALS FROM THE ATMOSPHERE TO THE OCEAN

### A.    NEARSHORE

There have been relatively few field studies of the atmospheric input of materials to coastal areas. Cambray et al. (1975) have attempted to measure the deposition of metals from the atmosphere into the North Sea by analyzing material collected in continuously open rain collectors exposed at several coastal sites for periods of approximately one month per sample. Samples collected in this manner over such a long period of time have a high probability of being contaminated by material from local sources; however, the investigators selected the sites carefully to minimize such contamination. On the basis of these measurements, integrated over a one-year period (1972–1973), Cambray *et al.* (1975) estimate that the annual atmospheric input of Fe, Pb, Zn, and Cu to the North Sea may be greater than 10 percent of the steady-state concentration of these metals in the water.

Patterson and Settle (1974) measured the atmospheric input of Pb into a 12,000 km² area of the Southern California Bight. They found that atmospheric transport accounted for 45 percent of the pollutant Pb input, the balance being carried by waste-water discharges, river input, and storm runoff.

These field studies clearly illustrate the potential importance of atmospheric deposition to nearshore marine pollution.

B. OPEN OCEANS

It is impossible at present to make any realistic estimate of the atmospheric transport of metals to the global ocean because of a complete lack of data from many regions, the most critical being the southern hemisphere. The data coverage is best in the North Atlantic, where there are several sets of data from different areas and covering relatively extended periods of time (see Table 5.1). With these data, mean deposition rates can be calculated on the basis of a simple model.

The model assumes that the metal-bearing aerosols are distributed uniformly from the sea surface to 5000 m. This distribution is based, by analogy, on measurements of a number of continentally derived non-pollutant elements (Gillette and Blifford, 1971). (There are virtually no data on metal concentrations at altitudes greater than a few meters above sea level.) The model further assumes that the atmosphere is washed clean of particles 40 times each year. This rain statistic is based on the fact that the average residence time of water vapor in the troposphere ranges from 8 days at low latitudes to about 15 days at high latitudes (Junge, 1963). Obviously, this crude calculation can yield only a rough estimate of the total deposition; however, the limitations of the data do not warrant a more sophisticated treatment.

The most representative data sets for the North Atlantic are those obtained by Duce *et al.* (1967b) for the period June through September 1973, from a 40-m tower at Bermuda, and by Chesselet *et al.* (1975) for April and May 1974, from a ship in an area of the eastern tropical

Atlantic between 18 and 36° N and 18 and 30° W (Table 5.1). Neither set of samples was collected in the plume of dust that periodically emerges from the Sahara and crosses the Atlantic in the northeast trade-wind belt. The eastern tropical Atlantic data were weighted twice as heavily as the Bermuda data because the former are more representative of the open Atlantic.

The weighted arithmetic means (Table 5.4) are used in making the flux calculations. Using atmospheric Fe as an example, we obtain the following input for Fe: 130 ng m$^{-3}$ STP × 40 washouts yr$^{-1}$ × 5000 m$^3$ STP m$^{-2}$ = 2.6 × 10$^{-2}$ g m$^{-2}$ yr$^{-1}$.

The calculated deposition rates of metals to the North Atlantic Ocean surface are given in column 3 of Table 5.4. How do these estimated atmospheric inputs compare with the input of these elements to this region from all sources? To answer this question, we estimate the total input on the basis of the known elemental composition of the deep-sea sediments and the measured sediment accumulation rates. (This calculation assumes, of course, that the oceans are in steady state.) The calculated sediment fluxes, based on a mean deep-sea clay sedimentation rate of 12 g m$^{-2}$ yr$^{-1}$ (Turekian, 1965), are given in column 4 of Table 5.4. The last column of Table 5.4 presents the

TABLE 5.4   Estimated Flux of Metals to the North Atlantic Ocean

| Metal | Atmospheric Concentration[a] (ng m$^{-3}$ STP) | Atmospheric Flux to Ocean (g m$^{-2}$ yr$^{-1}$) | Sediment Deposition[b] (g m$^{-2}$ yr$^{-1}$) | Atmospheric Input (% of Sed. Input) |
|---|---|---|---|---|
| Fe | 130 | 3 × 10$^{-2}$ | 7.8 × 10$^{-1}$ | 4 |
| Al | 200 | 4 × 10$^{-2}$ | 1.0 × 10$^{0}$ | 4 |
| Mn | 1.2 | 2 × 10$^{-4}$ | 8.0 × 10$^{-2}$ | 0.3 |
| Sc | 0.03 | 6 × 10$^{-6}$ | 2.3 × 10$^{-4}$ | 3.0 |
| Zn | 5 | 1 × 10$^{-3}$ | 2.0 × 10$^{-3}$ | 51 |
| Cu | 1.3 | 3 × 10$^{-4}$ | 3.0 × 10$^{-3}$ | 10 |
| Cd | 0.3 | 6 × 10$^{-5}$ | 5.0 × 10$^{-6}$ | 1200 |
| V | 0.6 | 1.2 × 10$^{-4}$ | 1.4 × 10$^{-3}$ | 8 |
| Sb | 0.12 | 2 × 10$^{-5}$ | 7 × 10$^{-6}$ | 300 |
| Pb | 7 | 1.4 × 10$^{-3}$ | 9.6 × 10$^{-4}$ | 150 |
| Hg | 0.10 | 2 × 10$^{-5}$ | 8.0 × 10$^{-6}$ | 250 |
| As | 0.1 | 2 × 10$^{-5}$ | 1.6 × 10$^{-4}$ | 13 |
| Se | 0.3 | 6 × 10$^{-5}$ | 2.0 × 10$^{-6}$ | 2900 |

[a] Weighted mean from data in Table 5.1 (see text).
[b] Calculated on the basis of deep-sea clay composition given by Turekian and Wedepohl (1961) as presented in *Geochemical Tables*, H. V. Rosler and H. Lange, eds. (Elsevier, New York, 1972), pp. 236–239, except for Hg (Bertine and Goldberg, 1971) and Sb (Duursma, 1973).

computed atmospheric input expressed as a percentage of the sediment deposition rate for each metal. The calculated percentage atmospheric inputs are subject to considerable uncertainty. Nonetheless, highly anomalous values can alert us to important processes or sources that might not otherwise be considered. In Table 5.4, the atmospheric input percentage values for Al, Fe, Cu, Mn, V, As, and Sc are reasonable. However, the atmospheric input rate for Zn is high, and those for Cd, Sb, Pb, Hg, and Se are greater than the deposition rates of these elements to the sediments. The elements in the high atmospheric deposition rate group are also among those that are anomalously enriched in the aerosol.

There are several possible explanations for the high atmospheric input rates for these elements:

1. The primary sources of these elements are anthropogenic, and, thus, the calculated input rates reflect a recent phenomenon in the geological time scale. The sedimentation data are long-term averages spanning many hundreds to thousands of years. Thus, an unusually high-rate, but short-term, injection would not be resolvable in the sediment record (see below).

2. A significant quantity of the metals present in the atmosphere may actually be recycled material that has been resuspended from the ocean surface. Most suspect would be those metals that are enriched in sea-salt aerosols relative to their seawater concentrations. Thus, the calculated atmospheric deposition rate would be artificially high and would not represent a true net input to the ocean.

3. The assumption that the mechanism for the removal of the metals from the atmosphere is equally efficient for all particle sizes may not be valid. Experiments indicate that aerosol removal by precipitation and by dry processes is relatively more efficient for larger particles (see Chapter 4). Thus, the actual removal rate for small particles over the oceans may be considerably less than that for large particles. As mentioned previously, measurements made on Bermuda (Duce *et al.*, 1976a) of particle composition as a function of size show that the major fraction of the mass of Cu, Zn, As, Sb Se, Hg, Pb, and Cd is present in submicrometer particles. In contrast, Na, Al, Mn, Fe, Sc, Th, and Co are found primarily on the 1- to 5-m-radius particles. Thus, it would seem that the actual deposition rates of the enriched metals could be much less than that predicted by our simple model. However, if the size-selective removal of aerosols were a major effect, we would expect to see a trend of progressively increasing enrichment factors with increasing distance from continental sources. No such trend is

evident in the data in Table 5.2, but this may be attributable to the limited nature of the data base.

At present, it is impossible to determine which of these, or other factors, might be responsible for the anomalously high calculated atmospheric input values. However, it must be emphasized that a proper understanding of the nature and magnitude of the flux to the oceans will require a careful investigation of the physical processes involved, in addition to the actual measurement of deposition on the ocean.

## C.  TIME RECORDS OF ATMOSPHERIC TRANSPORT TO THE OCEANS

The magnitude of anthropogenic inputs and the impact of these inputs on the global cycle of metals could be more readily assessed if we had available a chronological record of atmospheric deposition that predates the era of heavy industrialization. Such a record may be obtainable from marine sediments and glaciers.

### 1. Sediment Records

Deep-sea sediments cannot be used because of the low sedimentation rates that obtain, generally, a centimeter or less per thousand years. Thus, the anthropogenic components would be confined, theoretically, to a layer less than a millimeter thick at the sediment surface. In practice, burrowing benthic organisms churn the surface sediments to depths of at least several centimeters and, in effect, homogenize the time record over a several-thousand-year period. Nearshore sediments can yield much better time resolution because of the much higher sedimentation rates (Bruland et al., 1974). Burrowing organisms still pose a problem, but this can be avoided by working with sediments from anoxic basins (Bertine and Goldberg, 1977). However, the interpretation of nearshore sediment records is difficult, and the conclusions ambiguous, because of the complicating effects of inputs from rivers, sewage outfalls, and coastal runoff.

### 2. Glacial Records

There is only one source for the particulate and dissolved material incorporated into glaciers—the atmosphere. Also, the chronology of the annual inputs to glaciers is readily established by a number of techniques. Certainly, the transfer mechanisms and, consequently, the deposition rates to the glacier surface will differ from those over the ocean. However, a record of the year-to-year changes in the input rates of materials to the glacier are extremely valuable because they permit correlations to be made with known trends in (postulated) source processes such as the annual rates of fossil-fuel consumption, processing of specific raw materials, and volcanic activity.

A classic example is the study (Murozumi *et al.,* 1969) that showed that the concentration of Pb in the Greenland ice cap increased at a rate that reflected the increased industrial processing and utilization of the metal. This study was the first to demonstrate that the atmospheric transport of a metal was more than a local or regional phenomenon.

The glacial record will also provide information on the output from natural sources (such as crustal material, the ocean, and volcanoes), on the variation of these outputs with climate, and on the possible effects of output changes on climate.

However, glacier studies are fraught with difficulties, the major one being the problem of contamination during the sampling procedure. The contamination problem is severe because of the extremely low concentrations of particulate and dissolved materials in glaciers. The extreme precautions that must be taken, and the consequences of not taking them, are described by Murozumi *et al.* (1969).

## V. RECOMMENDATIONS

A. *Carefully designed field programs should be carried out in open ocean areas to measure the wet and dry deposition rates of metals to the sea surface.* Data are required for both the northern and southern hemispheres; *the need for data from the southern hemisphere is urgent.* Islands in midocean locations are preferred as sampling sites. Although sample collection will be difficult and contamination will continue to be a major problem, this type of study is presently feasible.

*Aerosol composition as a function of size is a critically important parameter in the determination of deposition rates;* the inclusion of such measurements into field programs should be encouraged.

B. *Investigations should be undertaken to determine the major sources of metals in the atmosphere.* Determination of sources and fluxes have equal priority. Emissions from anthropogenic sources such as smelters, cement plants, oil, and gas power plants must be measured and characterized on a global basis. *Important parameters are elemental (or chemical) composition, particle size distribution, and particle composition as a function of size.*

The importance of specific natural sources and processes such as volcanism, the biosphere, low-temperature volatilization and crustal degassing, and chemical fractionation at the air–sea interface must be evaluated.

C. *The time trends of the atmospheric transport of metals to remote areas should be established through studies of concentration changes with depth in glacier snow and ice.* With these measurements, and with a knowledge of the historical trends of anthropogenic source terms, an assessment can be made of the anthropogenic impact on a global scale. However, if glacier snow and ice is to be used for this purpose in a systematic fashion, *it will be necessary to develop new techniques for collecting larger cores using more stringent contamination-control procedures.*

D. *Immediate efforts must be made to initiate procedures or programs that will lead to an increase in the accuracy and precision of aerosol concentration and composition data.* There are two ways in which a given laboratory can verify the quality of its analytical procedures: (1) through the use of standard reference materials and (2) through participation in interlaboratory comparison exercises. Standard reference materials must be formulated according to the needs of environmental scientists working in the marine environment. At present, standards exist for metals in such solid phases as fish meal, orchard leaves, fly ash, and coal (U.S. National Bureau of Standards). Of these, only the latter two are suitable for use in some aerosol studies; however, they are of little value in the analysis of highly saline marine aerosols.

*It is essential that a mechanism be established whereby standards can be formulated and distributed expeditiously.* Often, developments in the area of environmental studies have occurred rapidly. In the past few years, there have been a number of instances where concern over the dissemination into the environment of certain elements or compounds resulted in a sharp increase in field sampling programs. In many cases, the quality of these data might have been greatly improved if standards could have been made available on short notice. At present, the U.S. National Bureau of Standards requires three years for the preparation of reference materials; this is far too slow a procedure for our purposes. Consequently, *we urge SCOR to explore the possibility of designating an appropriate international laboratory that would be responsible for the rapid preparation of and dissemination of environmental reference standards and for the organization of interlaboratory intercomparison exercises.*

# REFERENCES

Barger, W. R., and W. D. Garrett (1970). Surface active organic material in the atmosphere, *J. Geophys. Res. 75*, 4561–4566.

Barker, D. R., and H. Zeitlin (1972). Metal ion concentrations in sea-surface microlayer and size-separated atmospheric aerosol samples in Hawaii, *J. Geophys. Res. 77*, 5076–5086.

Beauford, W. J., J. Barker, and A. R. Barringer (1977). Release of particles containing metals from vegetation into the atmosphere, *Science 195*, 571–753.

Bertine, K. K., and E. D. Goldberg (1971). Fossil fuel combustion and the major sedimentary cycle, *Science 173*, 233–235.

Bertine, K. K., and E. D. Goldberg (1977). History of heavy metal pollution in Southern California coastal zone—reprise, *Environ. Sci. Technol. 11*, 297–299.

Bruland, K. W., K. Bertine, M. Liode, and E. D. Goldberg (1974). History of metal pollution in the Southern California coastal zone, *Environ. Sci. Technol. 8*, 425–431.

Buat-Menard, P., J. Morelli, and R. Chesselet (1974). Water soluble elements in atmospheric particulate matter over tropical and equatorial Atlantic, *J. Rech. Atmos. 8*, 661–673.

Cadle, R. D., A. F. Wartburg, W. H. Pollock, B. W. Ganrud, and J. P. Shedlovsky (1973). Trace constituents emitted to the atmosphere by Hawaiian volcanoes, *Chemosphere 2*, 231–234.

Cambray, R. S., D. F. Jeffries, and G. Topping (1975). *An Estimate of the Input of Atmospheric Trace Elements into the North Sea and the Clyde Sea (1972–3)*, AERE Rep. R 7733, United Kingdom Atomic Energy Authority, Harwell, 26 pp.

Chesselet, R., J. Morelli, and P. Buat-Menard (1972). Variations in ionic ratios between reference sea water and marine aerosols, *J. Geophys. Res. 77*, 5116–5131.

Chesselet, R., R. Buat-Menard, and M. Lesty (1975). Trace element chemistry in aerosols over the eastern torpical Atlantic, *Jean Charcot Cruise*, MIDLANTE Progress Rep., Numerical Data DFR/CNRS, Gif-sur-Yvette, France.

Chesselet, R., P. Buat-Menard, and C. Jehanno (1976). Trace metal associations and enrichments in oceanic microlayer derived aerosols assessed by scanning electron microscopy and the electron microprobe, presented at the American Meteorological Society, Second Conference on Ocean-Atmosphere Interactions, Seattle, Washington, March 30–April 1.

Chester, R., and S. H. Stoner (1974). The distribution of Mn, Fe, Cu, Ni, Co, Ga, Cr, V, Ba, Sr, Sn, Zn, and Pb in some soil-sized particulates from the lower troposphere over the world ocean, *Marine Chem. 2*, 157–188.

Crecelius, E. A., C. L. Johnson, and G. C. Hoffer (1974). Contamination of soils near a copper smelter by arsenic, antimony, and lead, *Water, Air, Soil Pollut. 3*, 337–342.

Crozat, G., J. L. Domergue, and V. Bogui (1973). Etude de l'aerosol atmospherique en Cote d'Ivoire et dans le Golfe de Guinee, *Atmos. Environ. 7*, 1103–1116.

Duce, R. A., and G. H. Hoffman (1976). Atmospheric vanadium transport to the ocean, *Atmos. Environ. 10*, 989–996.

Duce, R. A., G. L. Hoffman, and W. H. Zoller (1975). Atmospheric trace metals at remote northern and southern hemisphere sites: Pollution or natural? *Science 197*, 551–557.

Duce, R. A., B. J. Ray, G. L. Hoffman, and P. R. Walsh (1976a). Trace metal concentrations as a function of particle size in marine aerosols from Bermuda, *Geophys. Res. Lett. 3*, 339–343.

Duce, R. A., G. L. Hoffman, B. J. Ray, I. S. Fletcher, G. T. Wallace, J. L. Fasching, Sr.,

R. Piotrowicz, P. R. Walsh, E. J. Hoffman, J. M. Miller, and J. L. Heffter (1976b). Trace metals in the marine atmosphere: Sources and fluxes, in *Marine Pollutant Transfer*, H. Windom and R. Duce, eds., D. C. Heath & Co., Lexington, Mass. pp. 77–119.

Duursma, E. K. (1973). Specific activity of radionuclides sorbed by marine sediments in relation to the stable element composition, *Radioactive Contamination of the Marine Environment*, IAEA, Vienna, pp. 57–71.

Eriksson, E. (1959). The yearly circulation of chloride and sulfur in nature; meteorological, geochemical, and pedological implications, Part I, *Tellus 11*, 375–403.

Gillette, D. A., and I. H. Blifford (1971). Composition of tropospheric aerosols as a function of altitude, *J. Atmos. Sci. 28*, 1199–1210.

Goldberg, E. D. (1971). Atmospheric dust, the sedimentary cycle and man, *Comments in Geophysics: Earth Sci. 1*, 117–132.

Goldberg, E. D. (1976). Rock volatility and aerosol composition, *Nature 260*, 128–129.

Greenberg, R. R., W. H. Zoller, and G. E. Gordon (1977). Atmospheric toxic elements from refuse incineration. *Environ. Sci. Technol.*, to be published.

Hobbs, P. V., L. F. Radke, and J. L. Stith (1977). Eruptions of the St. Augustine volcano: Airborne measurements and observations, *Science 195*, 871–873.

Hoffman, G. L., and R. A. Duce (1972). Consideration of the chemical fractionation of alkali and alkaline earth metals in the Hawaiian marine atmosphere, *J. Geophys. Res. 77*, 5161–5169.

Hoffman, E. J., and R. A. Duce (1976). Factors influencing the organic carbon content of marine aerosols: A laboratory study, *J. Geophys. Res. 81*, 3667–3670.

Hoffman, G. L., R. A. Duce, and E. J. Hoffman (1972). Trace metals in the Hawaiian marine atmosphere, *J. Geophys. Res. 77*, 5322–5329.

Hoffman, E. J., G. L. Hoffman, and R. A. Duce (1974). Chemical fractionation of alkali and alkaline earth metals in atmospheric particulate matter over the North Atlantic, *J. Rech. Atmos. 8*, 675–688.

Johansson, T. B., R. E. Van Grieken, and J. W. Winchester (1974). Marine influences on aerosol composition in the coastal zone, *J. Rech. Atmos. 8*, 761–776.

Junge, C. E. (1963). *Air Chemistry and Radioactivity*, Academic Press, New York.

MacIntyre, F., and J. W. Winchester (1969). Phosphate ion enrichment in drops from breaking bubbles, *J. Phys. Chem. 73*, 2163–2169.

Maenhaut, W. R., W. H. Zoller, and R. A. Duce (1977). Atmospheric trace metals in Antarctica: Their source and transport, submitted to *J. Geophys. Res.*

Mroz, E., and W. H. Zoller (1975). Composition of atmospheric particulate matter from the eruption of Heimaey, Iceland, *Science 190*, 461–464.

Murozumi, N., T. J. Chow, and C. C. Patterson (1969). Chemical concentrations of pollutant lead aerosols, terrestrial dusts, and sea salts in Greenland and Antarctic snow strata, *Geochim. Cosmochim. Acta 33*, 1247–1294.

Newell, R. E., G. J. Boer, and J. W. Kidson (1974). An estimate of the interhemisphere transfer of carbon monoxide from tropical general circulation data, *Tellus 26*, 103–107.

Paciga, J. J., and R. E. Jervis (1976). Multielement size characterization of urban aerosols, *Environ. Sci. Technol. 10*, 1124–1128.

Patterson, C. C., and D. Settle (1974). Contribution of lead via aerosol deposition to the Southern California Bight, *J. Rech. Atmos 8*, 957–960.

Patterson, C. C., D. Settle, B. Schaule, and M. Burnett (1976). Transport of pollutant lead to the oceans and within the ocean ecosystems, in *Marine Pollutant Transfer*, H. Windom and R. Duce, eds., D. C. Heath & Co., Lexington, Mass., pp. 23–38.

Peirson, D. H., P. A. Cawse, and R. S. Cambray (1974). Chemical uniformity of airborne particulate material, and a maritime effect, *Nature 251*, 675–679.

Piotrowicz, S. R., B. J. Ray, G. L. Hoffman, and R. A. Duce (1972). Trace metal enrichment in the sea surface microlayer, *J. Geophys. Res. 77*, 5243–5254.

Rahn, K. A. (1976). The chemical composition of the atmospheric aerosol, Tech. Rep., Graduate School of Oceanography, U. of Rhode Island, Kingston, R.I.

Riley, J. P. (1975). Analytical chemistry of sea water, in *Chemical Oceanography*, 2nd ed., J. P. Riley and G. Skirrow, eds., Academic Press, London, pp. 193–513.

Robinson, E., and R. C. Robbins (1971). *Emission, Concentrations and Fate of Particulate Atmospheric Pollutants*, Publ. No. 4076, American Petroleum Institute, Washington, D.C., Final Rep., SRI Project SCC-8507.

Seiler, W. (1974). The cycle of atmospheric CO, *Tellus 26*, 116–135.

Seto, F. Y. B., and R. Duce (1972). A laboratory study of iodine enrichment on atmospheric sea salt particles produced by bubbles, *J. Geophys. Res. 77*, 5339–5349.

Study Panel on Assessing Potential Ocean Pollutants (1975). *Assessing Potential Ocean Pollutants*, a report to the Ocean Affairs Board, Commission on Natural Resources, National Research Council, Natural Academy of Sciences, Washington, D.C.

Sverdrup, G. M., K. T. Whitby, and W. E. Clark (1975). Characterization of California aerosols. II. Aerosol size distribution measurements in the Mojave Desert, *Atmos. Environ. 9*, 483–494.

Szekielda, K. H., S. L. Kupferman, V. Klemas, and D. F. Polis (1972). Element enrichment in organic films and foam associated with aquatic frontal systems, *J. Geophys. Res. 77*, 5278–5282.

Taylor, S. R. (1964). Abundance of chemical elements in the continental crust: a new table, *Geochim. Cosmochim. Acta 28*, 1273–1286.

Turekian, K. K. (1965). In *Chemical Oceanography*, J. P. Riley and G. Skirrow, eds., Academic Press, New York, Vol. 2, p. 106.

Van Grieken, R. E., T. B. Johansson, and J. W. Winchester (1974). Trace metal factionaters effects between sea water and aerosols from bubble bursting, *J. Rech. Atmos. 8*, 611–621.

Weiss, H. V., M. Koide, and E. D. Goldberg (1971). Mercury in a Greenland ice sheet: Evidence of recent input by man, *Science 174*, 692–694.

Whitby, K. T., W. E. Clark, V. A. Marple, G. M. Sverdrup, G. J. Sem, K. Willeke, B. Y. H. Liu, and D. Y. H. Pui (1975). Characterization of California aerosols. I. Size distributions of freeway aerosol, *Atmos. Environ. 9*, 463–482.

Wood, J. M. (1974). Biological cycles for toxic elements in the environment, *Science 183*, 1049–1052.

Young, J. A., N. S. Laulainen, L. L. Wendell, and T. M. Tanner (1975). The use of elemental concentration ratios to distinguish between the plumes of different northeastern cities, Ann. Rep., Battelle Northwest Laboratories, Richland, Wash.

Zoller, W. H., G. E. Gordon, E. S. Gladney, and A. G. Jones (1973). The sources and distribution of vanadium in the atmosphere, *Trace Elements in the Environment*, Advances in Chemistry Series No. 123, American Chemical Society, Washington, D.C. pp. 31–47.

Zoller, W. H., E. S. Gladney, and R. A. Duce (1974). Atmospheric concentrations and sources of trace metals at the South Pole, *Science 183*, 198–200.

# 6 Halogenated Hydrocarbons

## I. INTRODUCTION

The halogenated hydrocarbons have recently received considerable public attention because of the potentially detrimental impact of some species on the quality of our atmosphere and the earth's ecosystems (IMOS, 1975; Ocean Affairs Board, 1971). Of particular interest are data concerning the concentration distribution for the two classes of material that will be discussed in this chapter: the low-molecular-weight halocarbons (ICAS, 1975; Grimsrud and Rasmussen, 1975; Wilkniss *et al.*, 1975) and the high-molecular-weight chlorinated hydrocarbons (Bidleman *et al.*, 1976). Unlike most other pollutant materials that have substantial natural as well as anthropogenic sources, many of these species are believed to be exclusively man-made, and their transport to the ocean is recognized as an accomplished fact because it can be assumed that they were not present in seawater prior to their invention, manufacture, and dispersion by man. Although the assessment of impacts is not one of our objectives, we can state that the general concensus is that at this time there is no recognized direct impact of either group of materials on the ocean, although the bioaccumulation of the high-molecular-weight species has been established, and possible indirect effects via biomagnification in the food chain have been

---

Members of the Working Group on Halogenated Hydrocarbons were R. A. Rasmussen, *chairman*; T. F. Bidleman, D. Pierotti, and P. E. Wilkness.

postulated in specific cases. However, the data available on the horizontal and vertical concentration distributions in the oceans and in the atmosphere are very limited.

In this chapter, we will review our present knowledge on the distribution of these compounds in the atmosphere and the oceans and attempt to evaluate the most significant sources and transport paths.

## II. LOW-MOLECULAR-WEIGHT HALOCARBONS

In the evaluation of the tropospheric transport of the low-molecular-weight halocarbons, there are three categories that should be considered separately. One of these categories is based on source, and the other two are based on factors that affect tropospheric lifetimes: chemical reactions in the atmosphere; dissolution in, and removal by, precipitation; removal at land and ocean surfaces; removal after attachment to aerosol particles. The first category includes those compounds that appear to have their primary source in the ocean. The second category includes those man-made halocarbons whose tropospheric lifetimes are short, thereby limiting atmospheric transport to relatively short distances from the point of release. The third and final category includes those anthropogenic halocarbons whose tropospheric lifetimes are sufficiently long to allow transport over considerable distances through the troposphere; therefore these can be considered as potential global pollutants.

### A. OCEANIC SOURCE HALOCARBONS

In this group are the monohalomethanes $CH_3Cl$, $CH_3Br$, and $CH_3I$ (see Table 6.1 for a list of formulas, abbreviations, and names). These should not be considered as oceanic pollutants, because they are present in the ocean in concentrations far too high to be in equilibrium with the air above. (See Table 6.2 for a compilation of atmospheric concentration measurements.) Therefore, the ocean appears to be a source rather than a sink. Lovelock and co-workers (Lovelock *et al.*, 1973; Lovelock, 1975) have measured all three compounds in the Atlantic Ocean and have calculated that the ocean may release as much as $2.8 \times 10^{13}$ g yr$^{-1}$ of $CH_3Cl$ into the atmosphere. Wofsy *et al.* (1975) and Yung *et al.* (1975) have estimated the release of $CH_3Cl$ from the oceans at $5.2 \times 10^{12}$ g yr$^{-1}$, of $CH_3Br$ at $7.7 \times 10^{10}$ g yr$^{-1}$, and of $CH_3I$ at $7.4 \times 10^{11}$ g yr$^{-1}$. Liss and Slater (1974), using Lovelock's values for $CH_3I$, have calculated that the oceans release $2.7 \times 10^{11}$ g yr$^{-1}$ of this

TABLE 6.1    Formulas, Common Names, Abbreviations, and Their Equivalent Scientific Names

| | Scientific Name/Description |
|---|---|
| *Formula* | |
| $CH_3Cl$ | Methyl chloride |
| $CH_2Cl_2$ | Methylene chloride, dichloromethane |
| $CHCl_3$ | Chloroform |
| $CCl_4$ | Carbon tetrachloride |
| $CH_3Br$ | Methyl bromide |
| $CH_3I$ | Methyl iodide |
| $CH_3-CH_2Cl$ | Ethyl chloride |
| $CH_3-CCl_3$ | 1,1,1-Trichloroethane, methyl chloroform |
| $CHCl=CCl_2$ | Trichloroethylene |
| $CCl_2=CCl_2$ | Perchloroethylene, tetrachloroethylene |
| $CH_2=CHCl$ | Vinyl chloride |
| $CH_2Cl-CH_2Cl$ | Ethylene dichloride, 1,2-dichloroethane |
| $CH_2Br-CH_2Br$ | Ethylene dibromide, 1,2-dibromoethane |
| $CFCl_3$ | Freon-11 |
| $CF_2Cl_2$ | Freon-12 |
| $CHF_2Cl$ | Freon-22 |
| $CF_2Cl-CFCl_2$ | Freon-113 |
| $CF_2Cl-CF_2Cl$ | Freon-114 |
| *Common Name* | |
| Chlordane | *cis*- and *trans*- isomers of 1,2,4,5,6,7,8,8-octachloro-2,3,3a,4,7,7a-hexahydro-4,7-methanoindene |
| Dieldrin | 1,2,3,4,10,10-Hexachloro-*exo*-6,7-epoxy-1,4,4a,5,6,7,8,8a-octahydro-1,4-*endo*,*exo*-5,8-dimethanonaphthalene |
| Toxaphene | A mixture of chlorinated camphene compounds containing 67–69% Cl |
| *Abbreviation* | |
| BHC | Isomers of hexachlorocyclohexane |
| *p,p'*-DDD | 1,1-Dichloro-2,2-bis(*p*-chlorophenyl)ethane |
| *p,p'*-DDE | 1,1-Dichloro-2,2-bis(*p*-chlorophenyl)ethylene |
| *p,p'*-DDT | 1,1,1-Trichloro-2,2-bis(*p*-chlorophenyl)ethane |
| *o,p'*-DDT | 1,1,1-Trichloro-2-(*o*-chlorophenyl)-2-(*p*-chlorophenyl)ethane |
| PCB | Polychlorinated biphenyls |

compound, an estimate that is very close to that of McElroy (Wofsy *et al.*, 1975; Yung *et al.*, 1975), which was determined by a different method. The natural rates of production of these compounds in the oceans (and, possibly, on land by vegetation) far exceeds the rates from anthropogenic sources (see Table 6.3 for a summary of source production rates).

B. SHORT-RESIDENCE-TIME HALOCARBONS

The second class of compounds, those halocarbons with extremely short tropospheric lifetimes, includes $CHCl=CCl_2$, $CCl_2=CCl_2$, $CH_2Cl-CH_2Cl$, $CH_2=CHCl$, and $CH_2Br-CH_2Br$. The tropospheric lifetimes of these compounds are on the order of hours: $CH_2=CHCl$ (vinyl chloride), $CH_2Cl-CH_2Cl$ (ethylene dichloride), and $CHCl=CCl_2$ (trichloroethylene) have half-lives estimated at between 2 and 6 hours, and $CCl_2=CCl_2$ (perchloroethylene) has a half-life of around one day (Gay *et al.*, 1976). Consequently, if these compounds undergo tropospheric transport, they will have a very limited range. Nonetheless, Murray and Riley (1973) have measured $CHCl=CCl_2$ and $CCl_2=CCl_2$ in, and above, the waters of the Atlantic Ocean from the west coast of Africa to the British Isles. However, because $CHCl=CCl_2$, $CCl_2=CCl_2$, and $CH_2Cl-CH_2Cl$ all have been detected in municipal water and sewage systems in the United States, Great Britain, and Europe, much of the oceanic pollution may be transported through drainage waters rather than through the air. These compounds all have much longer lifetimes in water than in air (e.g., 6 months to a year for $CHCl=CCl_2$ and $CCl_2=CCl_2$), so it would seem that a significant portion of the oceanic burden could be due to direct injection by rivers. An alternative interpretation would be that Murray and Riley's data are evidence for the existence of natural sources for reduced halocarbons.

C. LONG-RESIDENCE-TIME HALOCARBONS

The chemical and physical properties of the low-molecular-weight $(C_1-C_2)$ halocarbons in this group are very different from those of the high-molecular-weight halocarbons, the polychlorinated biphenyls (PCB's) and chlorinated pesticides, which also have long residence times. (The latter group will be covered in Section III, below.) The low-molecular-weight halocarbons have much higher vapor pressures and, thus, tend to partition into the atmosphere. Their bioaccumulation factors are 100 to 1000 times smaller; and their water solubility, although fairly low, is considerably higher (i.e., $10^3$ to $10^5$ times as great). Also, the manner in which these compounds are used results in large releases to the atmosphere.

The chlorofluorocarbons, particularly Freon-11 (F-11), F-12, F-22, F-113, and F-114 (see Table 6.1) all appear to have long tropospheric lifetimes because of their relatively low reactivity in the troposphere and their low solubility coefficients in water. Therefore, we would expect the chlorofluorcarbons to have a sufficiently long residence

TABLE 6.2  Summary of Atmospheric Concentrations of Halomethanes[a] (Units: $10^{-12}$ v/v ± SD)

| Compound | Continental Background | Marine Background | Urban Range | Ratio, S to N Hemisphere[b] |
|---|---|---|---|---|
| *Nonfluorinated* | | | | |
| CH₃Cl | 569 ± 42[1]  530 ± 30[11] | 1260 ± 434[4] | 834 ± 40[4] | |
| | 587 ± 102[3] | 1140 ± 400[12] | | |
| | 713 ± 51[4] | | | |
| | 1040 ± 399[4] | | | |
| | 690 ± 390[6] | | | |
| CH₂Cl₂ | 36 ± 11[3] | 35 ± 19[6] | <20–144[3] | |
| CHCl₃ | 9 ± 3[1] | 40 ± 38[4] | 102 ± 102[4] | <0.11 |
| | 11 ± 5[3] | 27 ± 8[6] | 10–15,000[7] | |
| | 25 ± 18[4] | | 6–3000[9] | |
| | 20 ± 10[11] | | | |
| CCl₄ | 122 ± 13[1] | 128 ± 4[2] | 134 ± 20[4] | 0.95 |
| | 133 ± 10[3] | 128 ± 16[4] | 120–18,000[7] | |
| | 116 ± 8[4] | 111 ± 11[6] | 1400[8] | |
| | 120 ± 15[11] | | 120–1500[9] | |
| CH₃Br | 15 ± 10[4] | 93 ± 100[4] | 108 ± 138[4] | |
| | | | <10–220[5] | |
| CH₃I | 0.5 – 1.0[5] | 7 ± 7[4] | 24 ± 20[4] | |
| | 9 ± 5[4] | <1 – 11[10] | <1–3800[7] | |

| Compound | | | | |
|---|---|---|---|---|
| CH₃CCl₃ | 30.8 ± 3.7 | | | 0.42 |
| CCl₂CCl₂ | 30.7 ± 10.5[13] | | | 0.10 |
| CHClCCl₂ | 15.6 ± 2.5[13] | | | |

*Fluorinated*

| Compound | | | | |
|---|---|---|---|---|
| CFCl₃ (F-11) | 130 ± 5[1] | 123 ± 26[4] | 120–8800[7] | 0.89 |
| | 125 ± 8[11] | | 120–4900[9] | |
| | 115.6 ± 5.0[13] | | | |
| CF₂Cl₂ (F-12) | 228 ± 7[1] | 207 ± 33[4] | 250–47000[7] | 0.89 |
| | 230 ± 10[11] | | 250–11400[9] | |
| | 203.5 ± 18.5[13] | | | |
| CHCl₂F (F-21) | 14.2 ± 4.9[13] | | | |
| CFCl₂CF₂Cl (F-113) | 19.9 ± 3.4[13] | | | |

ᵃ Sources (complete citations are found in references):
1 Cronn et al. (1976).
2 Pierotti et al. (1976).
3 Pierotti and Rasmussen (1976).
4 Singh et al. (1977).
5 Harsch and Rasmussen (1977).
6 Cox et al. (1976).
7 Lillian et al. (1975).
8 Ohta et al. (1976).
9 Su and Goldberg (1976).
10 Lovelock et al. (1973).
11 Grimsrud and Rasmussen (1975).
12 Lovelock (1975).
13 Singh (1977).
ᵇ As reported in Singh (1977) from sources cited therein.

151

TABLE 6.3   Estimated Yearly Source Strengths of Chloroform and Three Monohalomethanes $(10^{12} \text{ g/yr})^a$

| | Methyl Chloride | Methyl Bromide | Methyl Iodide | Chloroform |
|---|---|---|---|---|
| Total source strength | 5.2 | $7.7 \times 10^{-2}$ | 0.74 0.27 | 0.23–0.99 |
| Global industrial source | $7.9 \times 10^{-3}$ | $(0.5–2.0) \times 10^{-2}$ | Insignificant | $12.4 \times 10^{-3}$ |

$^a$ Sources: Yung *et al.* (1975); Wofsy *et al.* (1975); Liss and Slater (1974).

time in the atmosphere to permit relatively complete tropospheric mixing on a northern–southern hemispheric scale. Indeed, the concentration of F-11 and F-12 has been found to be relatively uniform over nonurban continental regions and the oceans (Table 6.2). Because of the greater solubility in water of the low-molecular-weight halocarbons compared with DDT or the PCB's, substantial amounts would be expected to remain in solution if they were to be dumped into rivers or the ocean. Thus, they could become oceanic pollutants by this direct route in addition to the route through the atmosphere. Indeed, Thompson and co-workers (1975) have measured much higher concentrations of F-11 ($CFCl_3$) in municipal water systems than would be expected from equilibrium with F-11 in the atmosphere. Using Lovelock's measurements, Liss and Slater (1974) have calculated that the oceans act as a sink for F-11, removing $5.4 \times 10^9$ g yr$^{-1}$ from the atmosphere. However, this represents only about 2 percent of the total world production of F-11. Although specific measurements have not been made, a consideration of solubility coefficients relative to F-11 leads one to expect the ocean to be a greater sink for F-22 and a negligible sink for F-12, F-113, and F-114.

The low-molecular-weight chlorinated hydrocarbons with relatively long tropospheric lifetimes have much the same characteristics as the chlorofluorocarbons. Although generally they are more water-soluble than the chlorofluorocarbons, their solubility coefficients are, nonetheless, very small, and, consequently, precipitation scavenging is not an important mechanism for their removal from the troposphere. The compounds included in this group are $CCl_4$, $CHCl_3$, $CH_2Cl_2$, $CH_3CCl_3$, and $CH_3CH_2Cl$. $CCl_4$, $CHCl_3$, and $CH_3CCl_3$ all have been detected in, and over, the waters of the Atlantic Ocean by Lovelock *et al.* (1973) and by Murray and Riley (1973). The atmospheric concentrations of some of these compounds are summarized in Table 6.2, and the seawater concentrations of $CHCl_3$ and $CCl_4$ are presented in Table 6.4.

Liss and Slater (1974), using Lovelock's values, have calculated that the ocean acts as a sink for $CCl_4$, removing $1.4 \times 10^{10}$ g $yr^{-1}$ from the atmosphere. However, it must be reiterated that all these compounds are fairly soluble in water, and all (except $CH_3CH_2Cl$) have been detected in municipal water and sewage systems, along with F-11, F-113, and many other chlorinated compounds. Therefore, transport in the aqueous phase may be substantial; however, the primary route for transport to the ocean is most likely through the troposphere. The relative importance of these two modes could be assessed by comparing the concentrations of these compounds in estuarine and coastal waters with the concentrations in the open ocean, far from direct sources of injection from rivers and municipal sewage systems. This approach has been used for DDT and PCB's. Because most haloge-

TABLE 6.4   Chloroform and Carbon Tetrachloride Concentrations in Seawaters ($10^{-9}$ g $liter^{-1}$)

| Location | Chloroform | Carbon Tetrachloride | Reference |
|---|---|---|---|
| Open ocean, East Pacific (mixed layer, 0–100 m) Scripps Institution of Oceanography pier water, 1-28-75 to 7-8-75) | $14.8 \pm 5.3$ | $0.51 \pm 0.28$ | Su and Goldberg (1976) |
| Including rainy season | $11.8 \pm 5.8$ | $0.67 \pm 0.17$ | Su and Goldberg (1976) |
| Excluding rainy season | $9.3 \pm 3.6$ | $0.72 \pm 0.06$ | |
| Northeast Atlantic | $8.3 \pm 1.8$ | 0.17 | Murray and Riley (1973) |
| Liverpool Bay | — | 0.25[a] | Lovelock *et al.* (1973) |
| Atlantic Ocean | — | 0.41 | Pearson and McConnell (1975) |

[a] Includes tricholorethane ($CH_3CCl_3$).

nated hydrocarbons produced in the United States are manufactured along the Gulf Coast in Texas and Louisiana, the Gulf of Mexico would seem to be a prime area for making such measurements.

## II. REMOVAL OF HALOCARBONS FROM THE ATMOSPHERE

On the basis of the chemical properties of these materials and the emerging picture of their distribution patterns in nature, the following statements can be made:

1. For the major fraction of the low-molecular-weight halocarbons, chemical transformation is probably the most important removal mechanism, possibly with OH radical playing an important role (Singh, 1977).

2. The removal of low-molecular-weight halocarbons by precipitation does not appear to be very important; these compounds have solubility coefficients that are too small for this process to be efficient (Junge, 1976).

3. Gases are being directly transferred from the atmosphere to the oceans to a significant degree; this process is estimated to be of particular importance for such compounds as $CCl_4$, $CFCl_3$, and $CH_3I$ (Liss and Slater, 1974).

4. Attachment to, and subsequent removal by, aerosol particles is not likely to be an important mechanism in clean air for compounds having vapor pressures greater than $10^{-6}$ to $10^{-7}$ mm of Hg (Junge, 1975a). Thus, this process is insignificant for low-molecular-weight compounds but may be important for the high-molecular-weight chlorinated hydrocarbons (see below).

## III. CHLOROFLUOROCARBONS AS TRACERS OF AIR AND WATER MASSES

Freon-11 and Freon-12 ($CFCl_3$ and $CF_2Cl_2$, respectively) can be measured easily and with great sensitivity. Because of this, and because of their relative inertness in the troposphere, they could be ideal tracers with which to study the dispersion of pollutants from urban source areas and, subsequently, to follow atmospheric transport to the oceans. They may be especially useful for assessing the fate of reactive pollutant gases (e.g., CO, $SO_2$, and $NO_2$) in the same air mass; by simultaneously measuring pollutants along with F-11 and F-12, a correction can be made for reductions in concentration that result from dilution by mixing with clean air (i.e., with air parcels from above the boundary layer or from open-ocean trajectories). The usefulness of low-molecular-weight fluorocarbons for studying microscale and mesoscale transport phenomena has not been explored in practice, and knowledge in this area in general is essentially lacking. On the global scale, experimental observations show an increase in the atmospheric concentration of low-molecular-weight chlorofluorocarbons ($CFCl_3$ especially) with time; this trend is consistent with theoretical model predictions and worldwide production estimates (IMOS, 1975). The geographical distribution of these compounds (greater, and more variable, atmospheric concentrations in the northern hemisphere relative to the southern hemisphere and a gradient across the equator) has also been predicted by global modeling; this distribution can be attributed to the fact that these compounds are exclusively man-made and that their manufacture and use is concentrated in the northern hemisphere. In Table 6.2, the ratios of southern hemisphere to northern hemisphere concentrations are presented for some species; these ratios range from 0.10 for $CHClCCl_2$ to 0.95 for $CCl_4$.

The measurement of chlorofluorocarbons in seawater is relatively easy compared with the measurement of other tracers such as tritium or fission products. Only small volumes of seawater are required for analysis (less than 1 liter), making detailed investigations possible.

The concentration of $CFCl_3$ has been measured in the Atlantic (Lovelock *et al.*, 1973) and in the remote central Pacific. The measurements are too few and too fragmented to allow any detailed conclusions to be made on the sources, fluxes, or fate of this compound. A general observation that can be made is that the $CFCl_3$ is confined primarily to surface waters at this time and that its concentration decreases rapidly with depth. Thus it appears that $CFCl_3$ (and by analogy, the chlorofluorocarbons in general) are slowly being transferred to the oceans. The chlorofluorocarbons may be an extremely useful and unique tracer with which to follow mixing between surface water and the deep ocean.

However, the full potential of the chlorofluorocarbons as oceanographic tracers will not be realized until we have more, and better, data on the chemical fate of these compounds in seawater. Of particular interest are the rates of destruction through hydrolysis and through biodegradation. While data seem to indicate that the lifetime of these compounds in seawater is relatively long, experimental verification is necessary to assess the limits of these materials as oceanic tracers.

## IV. HIGH-MOLECULAR-WEIGHT CHLORINATED HYDROCARBONS

The second major class of halogenated hydrocarbons considered are the high-molecular-weight chlorinated hydrocarbons (CHC).

In 1971, the National Academy of Sciences published *Chlorinated Hydrocarbons in the Marine Environment*, a report that assessed the environmental impact and transport routes of these species (Ocean Affairs Board, 1971). In the same year, Woodwell *et al.* (1971) formulated a mass balance for DDT in the biosphere; the model incorporated translocation route from the land to the atmosphere to the ocean and predicted concentrations in these reservoirs through the end of the century. A similar model was proposed by Cramer (1973). These works emphasized the importance of thinking on a global scale and both concluded that, for these materials, the primary transport path to remote areas was via the atmosphere.

Since then, we have accumulated more information on the concentration of CHC in the air and the ocean, and it seems safe to conclude that DDT concentrations in these reservoirs are at least one to two orders of magnitude lower than predicted by Woodwell's global circulation model. Based on what has been learned since the early 1970's, new estimates of air–sea CHC fluxes can be made that are useful for two reasons: first, they allow us to assess whether atmospheric inputs could potentially account for the levels of CHC presently observed in the ocean; and, second, they suggest which atmospheric removal processes are likely to result in significant deposition into the oceans.

Although generally considered as being "nonvolatile," chlorinated pesticides and the PCB's have saturation vapor densities in the parts-per-million ($mg/m^3$) to parts-per-billion ($\mu g/m^3$) range. Thus, significant quantities of these materials can become airborne. The chief sources of these CHC's in the atmosphere are probably vaporization from soils for the chlorinated pesticides and evaporation of plasticizers and releases during open-dump burning for the PCB's (Nisbet and Sarofim, 1972).

Several laboratory studies have been carried out to measure pesticide evaporation rates. Some of these predict that the bulk of CHC pesticides applied to the soil would be lost to the atmosphere within a year. The few measurements of losses made under actual field conditions suggest that evaporation rates from soils are an order of magnitude lower than estimated from laboratory data (Willis *et al.*, 1972; Caro *et al.*, 1971; Caro and Taylor, 1971). There is a great need for more field studies if we are to formulate meaningful estimates of pesticide input to the atmosphere.

## V. CONCENTRATION AND DISTRIBUTION IN THE ATMOSPHERE

A large-scale survey of pesticides in U.S. air was carried out by the Environmental Protection Agency in 1970–1971 (Yobs *et al.*, 1972). A selection from these data (Table 6.5) shows that mean levels of airborne pesticides are only a few parts per trillion (ng/m$^3$), and even these means may be biased toward the high side as it appears from the authors' report that samples yielding values below the detection limit were not included in the averages. Other measurements of pesticides (Table 6.5) indicate that background levels over the continent are in the low ppt range and may be less than 1 ppt for DDT. This is substantially lower than Woodwell's predicted 72 ppt of DDT in the troposphere (Woodwell *et al.*, 1971) and more in line with Cramer's 0.11 ppt estimate (Cramer, 1973).

The mean concentration of the PCB's on suspended particles in four cities in the late 1960's was 50 ppm (Nisbet and Sarofim, 1972), which, based on the suspended particulate load, is equivalent to 3.5 ppt in the air. Measurements of airborne PCB's within the last two years are of similar magnitude (Table 6.5).

Direct evidence of pesticide transport over the oceans was provided in 1968–1971 by the identification of ppb quantities of DDT and dieldrin in trade winds aerosols collected on mesh samplers at Barbados (Risebrough *et al.*, 1968; Seba and Prospero, 1971; Prospero and Seba, 1972). Recent measurements over the North Atlantic (Table 6.6) indicate that nylon screen samplers have very likely underestimated the pesticide burden of marine air because most of the CHC is carried in the vapor phase rather than on particles (Bidleman *et al.*, 1976; Bidleman and Olney, 1974; Harvey and Steinhauer, 1974; Bidleman and Olney, 1975). The small number of available data indicate that CHC levels over the ocean are about 1 to 2 orders of magnitude lower than over land.

The concentration of CHC's in the atmosphere is not likely to decrease markedly in the near future. Although DDT has been banned in the United States and several northern European countries, the rate of use of DDT in South America, Africa, and Asia is expected to remain essentially constant or to increase slightly through the end of

TABLE 6.5 Chlorinated Hydrocarbons in Continental Air (Units: $10^{-9}$ g/m³ of air STP)

| Location | Year | PCB | DDT[a] | BHC | Chlordane | Dieldrin | Toxaphene | Reference |
|---|---|---|---|---|---|---|---|---|
| Mean of "average monthly levels," nine cities[b] | 1970–1971 | | 9.6 | 2.3 | | 2.3 | | Yobs et al. (1972) |
| New York, Florida, Texas | 1970–1971 | 80% of 200 samples contained CHC residues below 0.03–0.3 ppt | | | | | | Compton et al. (1972) |
| 760–2440 m above N.W. pine forest after DDT spraying | 1974 | | | | | | | Orgill et al. (1974) |
| | Immediately after spray | | 2.2–13.0 | | | | | |
| | 3 months after spray | | <0.02 | | | | | Orgill et al. (1974) |
| Mississippi Delta, mean of "monthly averages" | 1972 | | 99.5 | | | | 258 | |
| | 1973 | | 16.0 | | | | 82 | Arthur et al. (1976) |
| | 1974 | | 11.9 | | | | 159 | |
| Rhode Island | 1973–1975 | 1–15 | 0.03–0.6 | | 0.04–0.4 | | | Bidleman et al. (1976) |
| Sapelo Island, Georgia | 1975 | | 0.02–0.07 | | 0.1–0.3 | | 1.7–5.2 | Bidleman et al. (1976); Bidleman and Olney (1974) |
| Vineyard Sound, Massachusetts | 1973 | 4–5 | | | | | | Harvey and Steinhauer (1974) |
| La Jolla, California | 1973 | 0.5–14 | | | | | | McClure (1976) |
| London, UK | 1965 | | 3 | 6–11 | | 18–21 | | Abbott et al. (1966) |
| W. Germany | 1970 | | 0.3–0.6 | | | | | Junge (1975b) |
| W. Germany | 1970–1972 | | 0.01–1.0 | <0.01–0.1 | | <0.01–0.2 | | Weil et al. (1973) |

[a] May include p,p'-DDT, o,p'-DDT, and p,p'-DDE, depending on reference.
[b] Baltimore, Md.; Buffalo, N.Y.; Dothan, Ala.; Iowa City, Ia.; Orlando, Fla.; Salt Lake City, Utah; Fresno, Calif.; Riverside, Calif.; Stoneville, Miss.

158

TABLE 6.6  Chlorinated Hydrocarbons in Marine Air (Units: $10^{-9}$ m$^3$ of air STP)

| Location | Year | PCB | DDT[a] | Dieldrin | Chlordane | Toxaphene | Reference |
|---|---|---|---|---|---|---|---|
| Northeast tradewinds (particulate) | 1965–1966 | | 0.000061 | 0.000006 | | | Risebrough et al. (1968) |
| | 1968 | | 0.0002 | | | | Seba and Prospero (1971) |
| | 1970 | | 0.000066 | 0.000065 | | | Prospero and Seba (1972) |
| Bermuda | 1973 | 0.15–0.5 | 0.017–0.053 | | | | Harvey and Steinhauer (1974) |
| | 1973 | 0.19–0.66 | 0.009–0.022 | | <0.005–0.053 | <0.02–1.1 | Bidleman et al. (1976); Bidleman and Olney (1974); Bidleman and Olney (1975) |
| | 1974 | 0.08–0.48 | <0.003–0.062 | 0.003–0.077 | 0.01–0.067 | 0.11–1.9 | |
| Bermuda–U.S. cruises | 1973–1974 | 0.05–1.6 | 0.001–0.058 | 0.003–0.048 | <0.002–0.17 | 0.04–1.6 | Bidleman et al. (1976), Bidleman and Olney (1974); Bidleman and Olney (1975) |
| Grand Banks | 1973 | 0.05–0.16 | <0.001 | | | | Bidleman and Olney (1974) |
| Off British Coast | 1974 | <0.2–0.8 | 0.01–0.02 | | | | Dawsen and Riley (1977) |

[a] May include $p,p'$-DDT, $o,p'$-DDT, and $p,p'$-DDE.

159

TABLE 6.7  Chlorinated Hydrocarbons in Ocean Water (Units: $10^{-9}$ g/liter)

| Location | Year | PCB | DDT[a] | Reference |
|---|---|---|---|---|
| California coastal | 1971 | | 2–6 | Cox (1971) |
| waters | 1971 | 2.5 | 0.1 | Williams and Robertson (1975) |
| | 1973 | 0.3–0.5 | 0.3–1.3 | Young (1975) |
| | 1974 | 2–10 | 0.2–1.8 | Scura and McClure (1975) |
| | 1975 | <0.1–0.7 | <0.2–3.1 | Risebrough et al. (1976) |
| | 1975 | <0.1 | <0.05 | Risebrough et al. (1976) |
| Offshore waters of | | | | |
| Mexico | | | | |
| North Central | 1972 | 5 | <0.03 | Williams and Robertson (1975) |
| Pacific | | | | |
| North Atlantic | 1971 | 5–40 | | Bidleman et al. (1976); Olney and Wade (1972) |
| | 1972 | 1–22 | | Bidleman et al. (1976) |
| | 1972 | 27–41 | <1 | Harvey et al. (1973); Harvey et al. (1974) |
| | 1973 | 0.9–3.6 | <0.1 | Bidleman and Olney (1974) |
| | 1973 | 0.8–2.0 | | Harvey et al. (1974) |
| | 1975 | 4–9 | | Harvey and Steinhauer (1976) |
| Irish Sea | 1974 | <0.2–1.5 | <0.01–0.25 | Dawsen and Riley (1977) |
| Mediterranean, off | 1972 | 100–210 | 80–180 | Raybaud (1972) |
| Marseilles | | | | |
| N.W. Mediterranean | 1974–1975 | 1.5–38 | 0.5–2.7 | Elder (1977) |
| coastal waters | | | | |
| Open Mediterranean | 1975 | 0.2–8.6 | | Young (1977) |

[a] May include DDE and DDD, depending on reference.

160

the decade (Goldberg, 1975). Considering this geographical shift in DDT use, it is essential that measurements be made in the equatorial regions and in the southern hemisphere, as well as in the northern hemisphere.

## VI. CONCENTRATION AND DISTRIBUTION IN THE OCEANS

We have much more information on PCB and DDT residues in marine organisms than in seawater itself. Doubtlessly this is because seawater concentrations are very low, and contamination during sampling and analysis is a more serious problem than with animal tissues. Within the last few years, considerable progress in water-sampling techniques has been achieved by using macroreticular resins and other polymeric adsorbents to isolate trace organics from large volumes of water (Harvey *et al.*, 1973; Bedford, 1974; Junk *et al.*, 1974). Thus, the technology is now available to carry out a worldwide ocean-sampling program.

The data for CHC in seawater are summarized in Table 6.7. There is no completely satisfactory explanation for the apparent order-of-magnitude decrease in North Atlantic PCB concentrations between 1971 and 1973. Harvey attributed the decrease to the combined effects of scavenging by sinking particles and the declining input that resulted from world sales restrictions (Harvey *et al.*, 1974). However, it seems unlikely that sales restrictions would have such an immediate impact on PCB input—equivalent to "turning off the tap." Concentrations of PCB's in the North Pacific are comparable with those currently observed in the North Atlantic—a few ppt or less (Risebrough *et al.*, 1976; Scura and McClure, 1975; Williams and Robertson, 1975). No data exist for PCB's in southern hemisphere ocean waters.

Thus far, no one has reliably measured DDT in the open ocean, except in surface films (Bidleman and Olney, 1974). Concentrations in the subsurface water were below the detection limits of 0.1 ppt in the North Atlantic (Bidleman and Olney, 1974; Harvey *et al.*, 1974) and 0.03 ppt in the North Central Pacific (Williams and Robertson, 1975). Measurable concentrations of DDT have been reported off the California coast, ranging from 2 to 6 ppt in 1971 (Cox, 1971), 0.2 to 1.8 ppt in 1973–1974 (Young, 1975; Scura and McClure, 1975), and 0.05 to 0.8 ppt in 1975 (Risebrough *et al.*, 1976).

DDT and PCB's are enriched at the air–sea interface relative to their concentrations in the underlying water (Bidleman and Olney, 1974;

Bidleman *et al.*, 1976; Williams and Robertson, 1975; Olney and Wade, 1972; Duce *et al.*, 1972; Seba and Corcoran, 1969; Larsson *et al.*, 1974). The enrichment is a result of either scavenging by bubbles rising through the water column or atmospheric deposition on the sea surface. Lipids in surface films at the sea surface could accelerate the flux from the atmosphere to the ocean because the CHC's have much greater solubilities in lipids than in water. CHC's are probably removed from the water column by adsorption onto sinking particles (Harvey, 1974; Hom *et al.*, 1974), but nothing is known about the rate of this process in the open ocean.

## VII.  REMOVAL FROM THE ATMOSPHERE AND ATMOSPHERIC LIFETIMES

The presence of pesticides in rainwater and snow collected over land was the earliest indication of aerial CHC translocation. This led Woodwell *et al.* (1971) to consider rainfall the most important process for removing DDT from the atmosphere and prompted the Ocean Affairs Board (1971) to suggest that up to 25 percent of the annual DDT production would be rained out into the world's oceans. However, no data on CHC in rain collected at sea are available, and samples collected over land are likely to be locally contaminated. There is an urgent need for measurements of the concentration of CHC in precipitation at sea.

More CHC may be removed from the atmosphere through scavenging by falling particles than by rain, especially over land and on waters near the coast. In Sweden and Iceland, the annual rate of dry deposition for PCB's and DDT is of the order of grams per square kilometer (Sodergren, 1972; Bengston and Sodergren, 1974). McClure (1976) concluded that more CHC was deposited each year by dustfall than through washout. In California (Young, 1975; Young *et al.* 1976), aerial fallout contributed about as much DDT and one fourth the quantity of PCB's to the Southern California Bight as did surface runoff and sewage outfalls.

A potentially important, although as yet unexplored, mechanism of CHC input to the oceans may be vapor-phase transfer through the air–sea interface. Estimates of worldwide air–sea fluxes of several trace gases [$SO_2$, $N_2O$, $CO$, $CH_4$, $CCl_4$, $CCl_3F$, $CH_3I$, $(CH_3)_2S$] have been calculated by Liss and Slater (1974), assuming that the transfer occurs by molecular diffusion across stagnant air and water films (boundary layers) at the air–sea interface. Using this gas diffusion model and background levels at 0.1 ppt of PCB and 0.01 ppt of DDT over the western North Atlantic, Bidleman *et al.* (1976) have estimated maximum fluxes of $6 \times 10^7$ g yr$^{-1}$ of PCB and $6 \times 10^6$ g yr$^{-1}$ of DDT into the North American Basin (an area roughly $1 \times 10^7$ km$^2$ centered around the Bermuda Rise). Input estimates for rain were about equal to the vapor fluxes, and those from particle fallout, about half. The total PCB/DDT inputs into the North American Basin by atmospheric processes could be as large as $1.4 \times 10^8$ g yr$^{-1}$ of PCB and $1.4 \times 10^7$ g yr$^{-1}$ of DDT; this deposition is equivalent to about 0.3 percent and 0.013 percent of the 1971 world output of these chemicals, respectively. If these deposition rates were to be extrapolated to the world ocean (area $36.1 \times 10^7$ km$^2$), then the annual deposition rate, expressed as a percentage of 1971 production, would be 10.8 percent and 0.47 percent for PCB and DDT, respectively. There are few data to warrant such an extrapolation. However, we would expect the PCB and DDT concentrations to be greater over the North Atlantic relative to other ocean areas. Thus, the extrapolation yields an upper limit estimate that is considerably lower than earlier estimates.

While these estimates are only speculative, they do indicate that all three atmospheric removal processes are likely to be important—wet removal, dry deposition, and vapor-phase transfer. Furthermore, they are of the correct order of magnitude to account for the levels of the PCB's and DDT observed in the ocean surface waters. The residence time of these compounds in the 100-m ocean mixed layer could range from a year, as suggested by the 1972–1973 decline in North Atlantic surface water PCB concentrations, to as long as 10 years (about equal to the turnover rate of the water itself). The atmospheric input rate calculated above would thus result in mixed-layer concentrations ranging from 0.15 to 1.5 ppt of PCB and 0.015 to 0.15 ppt of DDT, values close to currently observed PCB levels in the North Atlantic and about equal to the present detection limit for DDT in seawater (Table 6.7). Certainly, for DDT, these estimates more nearly reflect the true situation in the ocean than do global mass balance predictions of 15 and 9 ppt in the mixed layer (Woodwell *et al.*, 1971; Cramer, 1973).

## VIII. RECOMMENDATIONS

1. *A major effort must be made to obtain halocarbon concentration data of the following types.*

    (a) *Vertical and areal distributions in the atmosphere.* Especially lacking are measurements of the high-molecular-weight chlorinated hydrocarbons in relatively remote ocean areas. The concentrations in such regions are very low, often below present-day detection limits. Thus, such measurements will require the development of better instrumentation and improved techniques to minimize the possibilities of losses and contamination. It is extremely important that measurements be made simultaneously of both the vapor-phase and the particulate-phase concentrations and of the concentration in precipitation.

*Most urgently needed are measurements in the southern hemisphere.* For some species such as the chlorofluorocarbons, the major centers of production and consumption presently lie in the northern hemisphere; for other species such as DDT, the major areas of consumption lie in the southern hemisphere. An extensive data base of concentrations in both hemispheres and a record of the changes in concentrations with time will be most useful for assessing fluxes and for validating atmospheric transport models.

(b) *Vertical and areal distributions in the oceans.* In order to assess the global fluxes to the ocean from the atmosphere and from rivers, the concentration fields in the surface (mixed) layer, and vertical profiles across the thermocline, must be well characterized. The concentration in phytoplankton and zooplankton should also be measured to determine the degree of biomagnification and the extent of natural biological production. For those compounds that have both natural and anthropogenic sources, the carbon-14 activity might be useful for determining the anthropogenic component and, therefore, should be measured. Intensive studies should be made in waters along the Gulf Coast, where many major manufacturers of halocarbons are situated.

2. *The modeling of the transport of halocarbons to the oceans calls for detailed knowledge of their chemical reaction mechanisms.* These can only be obtained through laboratory and field studies directed toward the identification of the important reactions and the dominant species. The importance of wide-ranging field studies cannot be stressed too highly. We anticipate that such studies will require the co-operative efforts of relatively large groups of chemists working in conjunction with meteorologists. From the standpoint of cost effectiveness, it may be desirable to attempt to integrate chemical programs into existing or planned large-scale meteorological experiments such as those that will take place as a part of the Global Atmospheric Research Program.

3. A record of the atmospheric concentrations of the halocarbons may reside in glaciers and snow fields. By comparing the preindustrial

concentrations of halocarbons in the lower layers of ice and snow cores with present-day concentrations in the upper levels of the cores, the relative strengths of natural and anthropogenic sources can be assessed.

4. At present, a wide variety of techniques are used to collect and identify halocarbons and to measure their concentrations. Often, the results obtained are not in agreement. Consequently, *we strongly recommend that a thorough intercalibration of these techniques be carried out as soon as possible.* It would be most useful to have a series of universally acceptable standards available for calibration purposes. Some group or organization should be assigned the responsibility for generating suitable standards.

## REFERENCES

Abbott, D. C., R. B. Harrison, J. O'G. Tatton, and J. Thompson (1966). Organochlorine pesticides in the atmosphere, *Nature 211*, 259.

Arthur, R., J. Cain, and B. Barrentine (1976). Atmospheric levels of pesticides in the Mississippi Delta, *Bull Environ. Contam. Toxicol. 15*, 129–134.

Bedford, J. (1974). The use of polyurethane foam plugs for extraction of PCB from natural waters, *Bull Environ. Contam. Toxicol. 12*, 622–625.

Bengtson, S. A., and A. Sodergren (1974). DDT and PCB residues in airborne fallout and animals in Iceland, *Ambio 3*, 84–86.

Bidleman, T. F., and C. E. Olney (1974). Chlorinated hydrocarbons in the Sargasso Sea atmosphere and surface water, *Science 183*, 516–518.

Bidleman, T. F., and C. E. Olney (1975). Long-range transport of toxaphene insecticide in the atmosphere of the western North Atlantic, *Nature 257*, 475.

Bidleman, T. F., C. P. Rice, and C. E. Olney (1976). High molecular weight chlorinated hydrocarbons in the air and sea. Rates and mechanisms of air/sea transfer, in *Marine Pollutant Transfer*, R. Duce and H. Windom, eds., D. C. Heath & Co., Lexington, Mass.

Caro, J. H., and A. W. Taylor (1971). Pathways of loss of dieldrin from soils under field conditions, *J. Agri. Food Chem. 19*, 379.

Caro, J. H., A. W. Taylor, and E. R. Lemon (1971). Measurement of pesticide concentration in air overlying a treated field, *Proceedings of International Symp. on Ident. and Meas. of Environ. Poll.*, Ottawa, Ont. Canada, Association of Official Analytical Chemists, Washington, D.C., p. 397.

Compton, B., P. Bazydlo, and G. Zweig (1972). Field evaluation of methods of collection and analysis of airborne pesticides, *Symposium on Pesticides in the Air*, 163rd ACS mtg., Boston, Mass.

Cox, J. L. (1971). DDT Residues in seawater and particulate matter in the California current system, *U.S. NOAA Fish. Bull. 69*, 433.

Cox, R. A., R. G. Derwent, A. E. J. Eggleton, and J. E. Lovelock (1976). Photochemical oxidation of halocarbons in the troposphere, *Atmos. Environ. 10*, 305–308.

Cramer, J. (1973). Model of the circulation of DDT on earth, *Atmos. Environ. 7*, 241.

Cronn, D. R., R. A. Rasmussen, and E. Robinson (1976). Report for Phase I of EPA Grant No. RO 804033-01, Aug. 23.

Dawsen, R., and J. P. Riley (1977). Chlorine-containing pesticides and PCB in British waters, *Estuarine Coastal Mar. Sci. 4*, 55–69.

Duce, R. A., J. G. Quinn, C. E. Olney, S. R. Piotrowicz, B. J. Ray, and T. L. Wade (1972). Enrichment of heavy metals and organic compounds in the surface microlayer of Narragansett Bay, R.I., *Science 176*, 161–163.

Elder, D. (1977). PCBs in N.W. Mediterranean coastal waters, *Mar. Pollut. Bull 7*, 63–64.

Gay, B. W., R. C. Noonan, and J. J. Bufalini (1976). Atmospheric oxidation of chlorinated ethylenes, *Environ. Sci. Technol. 10*, 58–67.

Goldberg, E. D. (1975). Synthetic organohalides in the sea, *Proc. R. Soc. Lond. B. 189*, 277–289.

Grimsrud, E. P., and R. A. Rasmussen (1975). Survey and analysis of halocarbons in the atmosphere by gas chromatography–mass spectrometry, *Atmos. Environ. 9*, 1014–1017.

Harsch, D. E., and R. A. Rasmussen (1977). Identification of methyl bromide urban air, submitted to *Environ. Sci. Technol.*

Harvey, G. R. (1974). In pollutant transfer to the environment. Deliberations and recommendations of the National Science Foundation, Pollutant Transfer Workshop, Port Aransas, Tex., Jan.

Harvey, G. R., and W. G. Steinahuer (1974). Atmospheric transport of polychlorobiphenyls to the North Atlantic, *Atmos. Environ. 8*, 777–782.

Harvey, G. R., W. G. Steinhauer, and J. M. Teal (1973). Polychlorobiphenyls in North Atlantic ocean water, *Science 180*, 643.

Harvey, G. R., W. G. Steinhauer, and H. Miklas (1974). Decline of PCB and DDT in the North Atlantic, in *Environmental Biogeochemistry*, J. O. Nrigau, ed., Ann Arbor Science Publ., Ann Arbor, Mich.

Hom, W., R. Risebrough, A. Soutar, and D. R. Young (1974). Deposition of DDE and PCB in dated sediments of the Santa Barbara Basin, *Science 184*, 1197–1199.

ICAS (1975). Interdepartmental Committee for Atmospheric Sciences (ICAS), *The Possible Impact of Fluorocarbons and Halocarbons on Ozone*, May 1975. (Available through Superintendant of Documents, U.S. Government Printing Office, Washington, D.C., Stock No. 03-000-00235-1.)

IMOS (1975). Report of Federal Task Force on Inadvertent Modification of the Stratosphere (IMOS), *Fluorocarbons and the Environment*, June, 109 pp. (Available through Superintendent of Documents, U.S. Government Printing Office, Washington, D.C., Stock No. 038-000-0226-1.)

Junge, C. E. (1975a). Basic considerations about trace constituents in the atmosphere as related to the fate of global pollutants, ACS Meeting, Philadelphia, Pa., April.

Junge, C. E. (1975b). Transport mechanisms for pesticides in the atmosphere, *J. Pure Appl. Chem. 42*, 95–104.

Junge, C. E. (1976). The role of the oceans as a sink for chlorofluoromethanes and similar compounds, *Z. Naturforsch. 31a*, 482–487.

Junk, G. A., J. J. Richard, M. D. Greiser, D. Witiak, J. L. Witiak, M. D. Arguello, R. Vick, H. J. Svec, J. S. Fritz, and G. V. Calder (1974). Use of macrorecticular resins in the analysis of water for trace organic contaminants, *J. Chromatog. 99*, 745–762.

Larsson, K., G. Odham, and A. Sodergran (1974). On lipid surface films on the sea. I. A simple method for sampling and studies of composition, *Marine Chem. 2*, 49–57.

Lillian, D., H. B. Singh, A. Appleby, L. Lobban, R. Arnts, R. Gumpert, *et al.* (1975). Atmospheric fates of halogenated compounds, *Environ. Sci. Technol. 9*, 1042–1048.

Liss, P. S., and P. G. Slater (1974). Flux of gases across the air–sea interface, *Nature 247*, 181–184.

Lovelock, J. E. (1975). Natural halocarbons in the air and in the sea, *Nature 256*, 193–194.

Lovelock, J. E., R. J. Maggs, and R. J. Wade (1973). Halogenated hydrocarbons in and over the Atlantic, *Nature 241*, 194–196.

McClure, V. E. (1976). Transport of heavy chlorinated hydrocarbons in the atmosphere, *Environ. Sci. Technol. 10*, 1223–1228.

Murray, A. J., and J. P. Riley (1973). Occurrence of some chlorinated aliphatic hydrocarbons in the environment, *Nature 242*, 37–38.

Ocean Affairs Board (1971). *Chlorinated Hydrocarbons in the Marine Environment*, National Academy of Sciences, Washington, D.C.

Nisbet, I. C. T., and A. F. Sarofim (1972). Rates and routes of transport of PCB in the environment, *Environ. Health Perspect. 1*, 21–37.

Ohta, T., M. Morita, and I. Misoguchi (1976). Local distribution of chlorinated hydrocarbons in the ambient air in Tokyo, *Atmos. Environ. 10*, 1–4.

Olney, C. E., and T. L. Wade (1972). Chlorinated hydrocarbon in the marine atmosphere and sea surface microlayer, IDOE Workshop on Baseline Measurement, Brookhaven, N.Y., May 22–26.

Orgill, M. M., M. R. Peterson, and G. A. Sehmel (1974). Some initial measurements of DDT resuspension and translocation from Pacific northwest forests, *1974 Symposium on the Atmosphere-Surface Exchange of Particulate and Gaseous Pollutants*, Richland, Wash.

Pearson, C. R., and G. McConnell (1975). Chlorinated $C_1$ and $C_2$ hydrocarbons in the marine environment, *Proc. R. Soc. Lond. 189*, 305–332.

Pierotti, D., and R. A. Rasmussen (1977). Interim Report for NASA Grant No. NSG 7214, Jan.

Pierotti, D., R. A. Rasmussen, J. Krasner, and B. Halter (1976). Trip report of the cruise of the R/V *Alpha Helix*, NSF Grant No. OCE 75 04688 A03, Sept.

Prospero, J. M., and D. B. Seba (1972). Some additional measurements of pesticides in the lower atmosphere of the northern equatorial Atlantic Ocean, *Atmos. Environ. 6*, 363–364.

Raybaud, H. (1972). Les biocides organochlores et les detergents anionique dans le milieu marin, Thesis, Centre Universitaire de Luminy, Universite d'Aix-Marseilles.

Risebrough, R., R. J. Huggett, J. J. Griffin, and E. D. Goldberg (1968). Pesticides: Transatlantic movements in the northeast trades, *Science 159*, 1233–1235.

Risebrough, R. W., B. W. deLappe, and W. Walker II (1976). Transfer of higher molecular weight chlorinated hydrocarbons to the marine environment, in *Marine Pollutant Transfer*, H. Windom and R. Duce, eds. D. C. Heath & Co., Lexington, Mass.

Seba, D. V., and E. F. Corcoran (1969). Surface slicks as concentrators of pesticides in the marine environment, *Pestic. Monit. J. 3*, 190–193.

Seba, D. V., and J. M. Prospero (1971). Pesticides in the lower atmosphere of the northern equatorial Atlantic Ocean, *Atmos. Environ. 5*, 1043–1050.

Scura, E., and V. McClure (1975). Chlorinated hydrocarbons in seawater: Analytical methods and levels in the northeastern Pacific, *Marine Chem. 3*, 337–346.

Singh, H. B. (1977). Atmospheric halocarbons: Evidence in favor of reduced average hydroxyl radical concentration in the troposphere, *Geophys. Res. Lett. 3*, 101–104.

Singh, H. B., L. Salas, A. Crawford, P. L. Hanst, and J. W. Spence (1977). Urban–nonurban relationships of halocarbons, $SF_6$, $N_2O$, and other atmospheric trace constituents, submitted to *Atmos. Environ.*

Sodergren, A. (1972). Chlorinated hydrocarbon residues in airborne fallout, *Nature 236*, 395–397.

Su, C., and E. D. Goldberg (1976). Environmental concentration and fluxes of some halocarbons, *Marine Pollutant Transfer*, Chap. 14, pp. 353–374, D. C. Heath and Co., Lexington, Mass.

Thompson, G. M., S. N. Davis, and J. M. Hayes (1975). Personal communication.

Weil, L., K. Quentin, and G. Ronicke (1973). Pestizidpegel des Luftstaubs in der Bundesrepublick, *Keinmission zur Erforschung der Luftverunreinigung*, Mitteilung VIII, Deutsche Forschungsgemeinschaft.

Wilkniss, P. E., J. W. Swinnerton, R. A. Lamontagne, and D. J. Bressan (1975). Trichlorofluoromethane in the troposphere, distribution and increase 1971–1974, *Science 187*, 832–833.

Williams, P. M., and K. J. Robertson (1975). Chlorinated hydrocarbons in sea-surface films and subsurface waters at nearshore stations and in the north central Pacific Gyre, *U.S. NOAA Fish. Bull. 73*, 445–447.

Willis, G. H., J. R. Parr, R. I. Papendick, and B. R. Carroll (1972). Volatilization of dieldrin from fallow soils as affected by different soil water regimes, *J. Environ. Qual. 1*, 193.

Wofsy, S. C., M. B. McElroy, and Y. L. Yung (1975). The chemistry of atmospheric bromine, *Geophys. Res. Lett. 2*, 215.

Woodwell, G. M., P. P. Craig, and H. A. Johnson (1971). DDT in the biosphere: Where does it go? *Science 174*, 1101–1107.

Yobs, A. R., J. A. Hanan, B. L. Stevenson, J. J. Boland, and H. F. Enos (1972). Levels of selected pesticides in ambient air of the U.S., *Symposium on Pesticides in the Air*, 163rd ACS Mtg., Boston, Mass.

Young, D. R. (1977). A synoptic survey of chlorinated hydrocarbon inputs to the Southern California Bight, Prog. Rep. to EPA, Grant No. R801153, Jan. 31.

Young, D. R., D.- J. McDermott, and T. C. Heesen (1976). Aerial fallout of DDT in Southern California, *Bull. Environ. Contam. Toxicol. 16*, 604–611.

Yung, Y. L., M. B. McElroy, and S. C. Worsy (1975). Atmospheric halocarbons: A discussion with emphasis on chloroform, *Geophys. Res. Lett. 2*, 397–399.

# 7 Petroleum and Related Natural Hydrocarbons

## I. INTRODUCTION

Petroleum is a complex mixture of chemical classes with many individual and physical properties and their atmospheric reactivity; these differences can affect the manner in which the materials are transported in the troposphere. The rate of hydrocarbon introduction into the atmosphere depends on the type of petroleum product, the volatility of its individual constituents, and, if used as a thermal energy source, the products produced in combustion. Also, there are a number of factors involved in the production, handling, transportation, and utilization of petroleum that affect its release into the atmosphere. In addition, the measurement of hydrocarbons in the atmosphere and the ocean and the assessment of anthropogenic impacts is complicated by the presence in the marine environment of a broad variety of biogenically produced hydrocarbons whose spectra would be superimposed on the petroleum hydrocarbon spectra.

Numerous studies have been made to determine the impacts of petroleum on the marine environment. These have been summarized recently in the report of a workshop on inputs, fates, and effects of petroleum in the marine environment (Ocean Affairs Board, 1975). In the report, the deleterious effects of high concentrations of petroleum from oil spills, petroleum handling, and ship operations are assessed.

Members of the Working Group on Petroleum and Related Natural Hydrocarbons were W. D. Garrett, *chairman*; J. H. Hahn, and J. W. Swinnerton.

These include amenity reduction, damage to recreational values, seabird mortality, tainting of seafood, and the modification of benthic communities in heavily polluted coastal sediments. Also, there are a number of possible chemical and physical mechanisms by which petroleum hydrocarbons could produce undesirable effects. Large petroleum slicks could change the ocean's albedo, alter the rate of exchange of gases through the air–water interface, and accumulate lipophilic substances. At this time, there is no evidence that petroleum films have had any significant impact on interfacial properties and processes. Theoretically, all of these mechanisms are plausible. However, no experiments have been performed to ascertain if these mechanisms are operative *in situ*. Indeed, information as basic as the distribution and areal extent of oil slicks on the oceans is lacking.

## II. ATMOSPHERIC CONCENTRATIONS AND AIR–SEA EXCHANGE

The existing data on the sources, concentration, and composition of hydrocarbons in the atmosphere is sparse, confusing, and often contradictory. Robinson and Robbins (1968) calculated that worldwide anthropogenic emissions of all types of hydrocarbons to the atmosphere resulted in a nonmethane hydrocarbon production of 68 Tga (Tga, teragrams per annum = $10^{12}$ g yr$^{-1}$). The authors estimate that about 30 Tga (of the 68 Tga released) are "reactive," i.e., they are largely olefinic, and the oxides of nitrogen will react rapidly (within hours) in the presence of sunlight and $NO_x$, ultimately forming condensed particles that have an atmospheric residence time of only a few days. The balance of 38 Tga remains in the gas phase; this material is referred to as the "long-lived" hydrocarbon fraction. On the other hand, Hidy and Brock (1970) estimated that only about 3 Tga of the hydrocarbons are rapidly converted to atmospheric particulate matter, with 65 Tga remaining in the gaseous form. An assessment by Peterson and Junge (1971) yielded 45 Tga of anthropogenic nonmethane hydrocarbons, of which 30 Tga remain as a gas and 15 Tga undergo conversion to particles. Table 7.1 (Duce et al., 1974a; 1974b) presents a summary of estimates of the total anthropogenic production of atmospheric hydrocarbons (excluding methane) that may largely remain in the gas phase and, thus, have relatively long residence times. Duce et al. (1974a; 1974b) concluded that the total man-made "long-lived" gaseous hydrocarbon production is approximately 50 Tga, of which 45 Tga is produced in the northern hemisphere.

TABLE 7.1   Estimates of Emissions and Conversion of Nonmethane Hydrocarbons in the Atmosphere

| Author | Total Emissions to Atmosphere (Tga)[a] | Rapid Conversion to Particles (Tga)[a] | Remaining as "Long-Lived" Gaseous Hydrocarbon (Tga)[a] |
|---|---|---|---|
| *Man's production* | | | |
| Robinson and Robbins (1968) | 68 | 30 | 38 |
| Hidy and Brock (1970) | 68 | 3 | 65 |
| Peterson and Junge (1971) | 45 | 15 | 30 |
| *Natural production* | | | |
| Rasmussen and Went (1965) | 440 | | |
| On the basis of | | | |
| Peterson and Junge (1971) | | 220 | 220 |
| Rasmussen (1973) | | ≤440 | ≈0 |

[a] Tga = Teragrams per annum = $10^{12}$ g year$^{-1}$.

The natural production of atmospheric hydrocarbons (mostly terpenoids) has been estimated to be as much as 440 Tga worldwide (Rasmussen and Went, 1965; Rasmussen, 1972; Peterson and Junge, 1971; Ripperton *et al.*, 1967). Peterson and Junge assumed that half of this material is converted relatively rapidly to particles, the balance remaining in the gaseous state. In contrast, Rasmussen suggested that almost all of the naturally produced hydrocarbons are rapidly converted to oxygenated species such as alcohols, aldehydes, and acids. Eschenroeder (1974) surveyed the state of knowledge of the sources of naturally emitted hydrocarbons in the atmosphere and concluded that the emission rate of natural hydrocarbons is offset by their reaction rate in the atmosphere. Thus, the estimates of the production rate of naturally produced hydrocarbons having relatively long atmospheric residence times range from 0 to 220 Tga (see Table 7.1). If it is assumed that production is proportional to land surface area, the production rate of "long-lived" gaseous hydrocarbons in the northern hemisphere would be 0 to 150 Tga.

A recent workshop (Ocean Affairs Board, 1975) estimated the global atmospheric input of petroleum hydrocarbons to the seas to be roughly 0.6 Tga. Duce (1973) estimated a higher value of 1.4 Tga for the heavier particulate hydrocarbons ($n$-$^{14}$C to $n$-$^{33}$C) carried to the sea by dry fallout and precipitation processes; these values are based on preliminary measurements of particulate hydrocarbons in air samples taken aboard a ship in the open North Atlantic Ocean. The cited estimate of 1.4 Tga was calculated from an average particulate hydrocarbon concentration of about $10^{-9}$ g/m$^3$ in the marine atmosphere; the hydrocarbons in the gas phase were not measured. Because these estimates are based on many unproven assumptions, the hydrocarbon input values should be regarded with caution.

A significant fraction of the nonmethane petroleum hydrocarbons in the marine atmosphere may be derived from oceanic sources. Garrett and Smagin (1976) estimate that 1.35 Tga is emitted to the atmosphere by natural seeps, transportation sources, and offshore production facilities. The majority of this sea-to-air petroleum flux is localized along tanker routes and coastal areas, where petroleum production activity is high. It is merely fortuitous that the sea-to-air flux is essentially equal to Duce's (1973) air-to-sea flux of 1.4 Tga, as the composition of the materials in these fluxes is quite different. The hydrocarbons entering the atmosphere from maritime petroleum sources will be primarily composed of volatile, gas-phase compounds that are highly paraffinic and stable and, thus, not likely to be rapidly converted into particles that would then be redeposited to the sea. A number of investigators have demonstrated experimentally that $n$-alkanes with up to 14 carbon atoms are rapidly evaporated from the sea surface, while McAuliffe (1973) suggests that most hydrocarbons with up to 12 carbon atoms (alkanes, alkenes, cycloalkanes) will be preferentially distributed into the gas phase under equilibrium conditions between air and sea. In contrast, Duce's downward flux consists of particulate matter formed from the heavier organic constituents of petroleum and their reaction products. Thus, in addition to continental inputs, it seems likely that there is a cycle of petroleum hydrocarbons between the oceans and the marine atmosphere in which (a) there is a net flux of volatile, low-molecular-weight compounds from sea to air; (b) the less stable compounds undergo gas-to-particle conversion in the atmosphere; and (c) the particulate forms are returned to the sea by precipitation processes and as dry fallout.

## III. FACTORS AFFECTING THE DETERMINATION OF ATMOSPHERIC TRANSPORT OF HYDROCARBONS

### A. INTRODUCTION

Estimates of atmospheric inputs of hydrocarbons to the ocean are very uncertain because of the paucity of data and the necessity of making a large number of simplifying assumptions when dealing with the complex physical and chemical systems inherent in a multicomponent mixture such as petroleum in the environment. There are five major problem areas that must be considered in attempting to estimate the magnitude of the tropospheric transport of petroleum hydrocarbons to the sea. These are related to (1) the complex nature of petroleum, (2) selective reactivity in the atmosphere, (3) selective transport, (4) natural sources, and (5) concentration levels.

## B.   THE COMPLEX NATURE OF PETROLEUM

Crude oil is composed of thousands of individual compounds, mostly hydrocarbons representing several homologous series. The relative ratios of these compounds in a petroleum product will determine the character of its combustion products when burned and its physical and chemical behavior when spilled onto water or allowed to evaporate. Although there are widespread variations in the constituents of crude oil from different sources, the following composition is representative of an "average" crude.

*By molecular size*: gasoline ($^5C - ^{10}C$), 30 percent; kerosene ($^{10}C - ^{12}C$), 10 percent; light distillate oil ($^{12}C - ^{20}C$), 15 percent; heavy distillate oil ($^{20}C - ^{40}C$), 25 percent; residual oil ($^{40}C$), 20 percent.

*By molecular type*: paraffin hydrocarbons (alkanes), 30 percent; naphthene hydrocarbons (cycloalkanes), 50 percent; aromatic hydrocarbons, 15 percent; nitrogen-, sulfur-, and oxygen-containing compounds (NSO's), 5 percent.

## C.   SELECTIVE REACTIVITY IN THE ATMOSPHERE

Once organic matter is injected into the atmosphere, its character will be greatly altered because of the large number of photochemical reactions that can take place. Compounds may be simply oxidized, or they may undergo complex polymerization reactions leading to particle formation. Strictly speaking, once oxidized, a compound can no longer be considered a hydrocarbon. The product is more polar and surface-active and is readily removed from the atmosphere by physical scavenging processes. The reactivity of airborne hydrocarbons in the presence of sunlight and oxidants increases in the following order: normal paraffins, branched-chain paraffins, aromatics, and olefins. Because of the variability in the chemical composition of petroleum, the differences in the reactivity of its components in the atmosphere, and the lack of sufficient kinetic data, it is difficult to make a realistic estimate of hydrocarbon losses from reactions in the atmosphere.

Actual measurements will be required to obtain data for use in transport models.

## D. SELECTIVE TRANSPORT

The magnitude of the transport of specific hydrocarbons to the sea will depend to a considerable extent on their physical state and water solubility. In a recent paper, Junge (1975) discussed various processes that might effect the removal of trace substances from the atmosphere. He concluded that the most effective removal processes should be dry deposition and wet removal, the latter as a consequence of the incorporation of particulate matter into raindrops as condensation nuclei. However, there are virtually no data on the types and concentrations of hydrocarbons in rainwater in remote marine areas (or even in nonurban continental regions); also, the air–sea exchange of hydrocarbons is poorly understood. Therefore, we have no basis for making any useful estimates.

Duce *et al.* (1974a; 1974b) concluded that a considerable fraction of the "long-lived" gaseous hydrocarbons found in the atmosphere in remote areas is probably not deposited on the earth's surface but is ultimately converted in the atmosphere to oxygenated species. This might be true, but the extent to which this conversion occurs will be open to question until measurements are made of the magnitude and character of the "long-lived" hydrocarbons attached to aerosol particles that are removed by dry fallout and by rainout and washout.

Until now, the only (and still rather uncertain) assessment of the tropospheric residence time of gaseous hydrocarbons has been made by Duce *et al.* (1974a; 1974b). They estimated a mean value of 0.5 to 2.0 years for the northern hemisphere for all hydrocarbons that are not converted, or attached, to particles; this estimate assumes a total northern hemisphere tropospheric burden of 91 Tg and anthropogenic and natural productions of 45 and 150 Tga, respectively.

If the continental regions are important sources of petroleum that is transported to the sea, one would expect decreasing concentrations of particulate and soluble forms of hydrocarbons with increasing distance from land. In addition, atmospheric concentrations should be greater in air masses that have passed over strong petroleum sources. These speculations have yet to be verified by measurement. However, indirect evidence suggests that concentrations are markedly greater close to the continents. This can be inferred from the Aitken particle concentration distribution field over the oceans (Hogan *et al.*, 1973; Elliot *et al.*, 1974) and from the frequency of occurrence of haze at sea

(McDonald, 1938). Aitken and haze particles are submicrometer sized and are formed predominantly by condensation from the gaseous state; thus, it can be assumed that the major portion of the airborne organic material resides in these particle classes. The highest concentrations at sea are found near the continents within 500 to 1000 km of the coast (Hogan, 1975; Machta, 1975; Prospero, 1975).

E.  NATURAL SOURCES OF HYDROCARBONS

The contribution of biogenically produced hydrocarbons to the total budget of hydrocarbons in the marine atmosphere is essentially unknown. The biogenic hydrocarbons could be derived from continental sources or from the sea itself. As far as continental sources are concerned, it is known that large quantities of terpenes are produced by vegetation, but these compounds are highly reactive, as mentioned above, and probably do not remain in true hydrocarbon form for a sufficient period of time to permit a significant transport to the sea via the troposphere.

The oceans may be a significant source of low-molecular-weight hydrocarbons. Measurements by Linnenbom and Lamontagne during the past nine years have yielded the following average open-ocean concentrations: ethane, 0.50 ppbv (ppbv = $10^{-9}$ by volume); ethylene, 4.8 ppbv; propane, 0.34 ppbv; and propylene, 1.4 ppbv. Nearshore samples usually contain 100 to 200 times the open-ocean concentrations of ethane and propane. The concentrations of ethylene and propylene appear to be related to biological activity and productivity (Lamontagne *et al.*, 1974; Swinnerton and Linnenbom, 1967; Swinnerton and Lamontagne, 1974). Vertical profiles in the ocean reveal that the $^{1}C-^{4}C$ concentrations reach a maximum at about 75 m; all but methane decrease rapidly to zero concentration below 75-m depth. The occurrence of these maxima in the open ocean indicates the existence of sources that are producing material at rates that are fast relative to the rates of dispersion due to physical mixing. The phytoplankton and chlorophyll *a* distributions in the upper layers also show pronounced maxima that often coincide with the hydrocarbon maxima; this suggests a relationship between hydrocarbon distributions and biological processes.

According to a review in the report of the workshop held by the Ocean Affairs Board (1975), almost every hydrocarbon family (even polynuclear aromatics) are produced somewhere in the biosphere. However, the relative concentrations of the predominant biogenically produced compounds differ significantly from those petroleum hydrocarbons. Criteria for discriminating between biogenic and petroleum hydrocarbons in sea samples are reviewed in the report of the workshop held by the Ocean Affairs Board (1975) along with appropriate sampling and analytical methodology. It is stated that hydrocarbon-class separation with liquid chromatography followed by gas-chromatographic analysis can make the desired distinction as long as the petroleum fraction is greater than 10–20 percent of the total hydrocarbons and if the quantity of material sampled is sufficiently large so that limits of detection are not approached. With small sample quantities, one must utilize more sophisticated methods—computerized gas chromatography and mass spectroscopy.

## F. CONCENTRATION LEVELS IN THE ATMOSPHERE

The most common hydrocarbons in both urban and rural air are the paraffins. Among the paraffins, considering the full carbon range from $^2C$ to $^{30}C$, the most prominent species appear to be the n-alkanes. For the urban situation, the major individual compounds in the range from $^6C$ to $^{12}C$ were found to be benzene homologues. In urban air, the total concentration of paraffins (excluding methane) was found to be in the range 200–300 g/scm (standard cubic meter) of air (Stephens and Burleson, 1967; USDHEW, 1970; Lonneman et al., 1974; Kopczynski et al., 1975). In rural areas, the concentration of nonmethane paraffins is lower by a factor of about 10, 15–30 g/scm of air (Ketseridis et al., 1976).

For long-range transport and deposition in the oceans, only the "long-lived" fraction of the atmospheric hydrocarbons (i.e., the aliphatic hydrocarbons, acetylene, and some of the aromatic hydrocarbons) is of interest. Although a few measurements of atmospheric concentrations of nonmethane hydrocarbons have recently become available, virtually no knowledge of the atmospheric distribution of these compounds is at hand. Measurements by Rasmussen (1974) at Hawaii and on cruise no. 32 of the *Meteor* from northern Europe to Puerto Rico suggest that the average atmospheric concentration of gaseous $^3C$–$^{12}C$ hydrocarbons is about 10 $\mu g$/scm in uncontaminated marine air. Duce et al. reported preliminary data on the atmospheric concentration of heavier hydrocarbons ($^{14}C$–$^{32}C$) measured from a tower in Bermuda from May to December 1973. The results indicated that about 95 percent of the hydrocarbons in this range are gaseous; they are not retained by Gelman type A glass fiber filters but are retained on polyurethane plugs (Bidleman and Olney, 1974) placed behind the filters. The concentration of the heavier hydrocarbons ranged up to about 1 g/scm of air. Twenty to twenty-five percent of this material was found to consist of gas-chromatographically resolvable components, while the remaining 75–80 percent was a complex mixture of unresolved components.

In a recent paper, Junge (1975) tentatively formulated a relationship between the surface area of aerosol particles, the vapor pressure of atmospheric trace constituents, and the fraction of the total atmospheric concentration of a trace compound that is attached to aerosol particles. As Junge pointed out, the relationship can only be used to obtain order-of-magnitude estimates; however, the relationship yields results that are in good agreement with the results of Duce *et al.*, since it predicts that, in the range of the $^{14}C-^{28}C$ normal alkanes, a very great portion of the compounds present in the atmosphere should be in the gas phase. In samples of organic particulate matter collected on glass fiber filters in remote regions such as the southwest coast of Ireland, the Cape Verde Islands, and the central Atlantic on *Meteor* cruise 32, October–December 1973, Ketseridis *et al.* (1976) identified normal alkanes with carbon numbers from 10 to 28 and a number of polyaromatic hydrocarbons up to chrysene and benzo(a)anthracene. Also, there was some evidence for the occurrence of pristane and phytane and a major amount of branched-chain alkanes. The data of Ketseridis *et al.* indicate that, with respect to the organic constituents, there are two different types of airborne particles in marine air: (1) a "young" aerosol comprised of organic material that is relatively unmodified by reactions in the atmosphere and that has a relatively high concentration of ether-extractable organics (5–15 percent of the total particulate matter) and (2) an "aged" aerosol composed of matter highly altered by reactions in the atmosphere and having a distinctly lower concentration of ether extractables (1–2 percent). The "young" aerosol matter collected in the central Atlantic was found to contain more than 1.4 percent (0.25 $\mu$g/scm of air) alkanes, whereas the "aged" matter collected in the same region consisted of 0.2–0.3 percent (30–60 ng/scm of air) alkanes.

The existing data, though meager, can serve as a basis for making a rough estimate of the total concentration of nonmethane hydrocarbons in remote areas (i.e., areas relatively far removed from strong anthropogenic sources). Estimates for the marine and for the continental atmosphere are presented in Table 7.2. The table is primarily based on a summary given by Duce *et al.* (1974a; 1974b) with the addition of recent data from Ketseridis *et al.* (1976). The latter study focused on essentially the same range of *n*-alkanes as that measured by Quinn and Wade (as reported in Duce *et al.*, 1974a; 1974b); the agreement be-

TABLE 7.2    Major Nonmethane Gaseous and Particulate Hydrocarbons in the Atmosphere of the Northern Hemisphere

| Hydrocarbons | Concentration Range ($\mu$g/scm air) | Reference |
|---|---|---|
| *Marine atmosphere* | | |
| $n$-$^3$C to $n$-$^{12}$C | 5–20 | Rasmussen (1974) |
| $n$-$^{14}$C to $n$-$^{32}$C | 0.2–1 | Quinn and Wade (unpublished data)[a] |
| $n$-$^{10}$C to $n$-$^{28}$C | 0.9 | Ketseridis *et al.* (1976) |
| *Continental atmosphere* | | |
| $n$-$^2$C to $n$-$^{12}$C | 50–200 | Rasmussen (1974) |
| $n$-$^{14}$C to $n$-$^{32}$C | 1–7 | Quinn and Wade (unpublished data)[a] |
| $n$-$^{10}$C to $n$-$^{28}$C | 10 | Ketseridis *et al.* (1976) |

[a] As reported in Duce *et al.* (1974).

tween the two data sets is reasonably good. On the basis of these
data, Duce *et al.* estimated the northern hemisphere tropospheric
burden to be about 11 Tg over the oceans and about 80 Tg over the
continents.

## IV.    RECOMMENDATIONS

The foregoing presentation and the recommendations that follow parallel those incorporated in a Technical Report of the World Meteorological Organization, *Determination of the Atmospheric Contribution of
Petroleum Hydrocarbons to the Oceans* (Garrett and Smagin, 1976).

1. The assessment of the atmospheric input of petroleum hydrocarbons will require measurements of the concentrations of hydrocarbons
in the atmosphere, at the air–sea interface, and in precipitation and dry
fallout. Emphasis should be placed on the hydrocarbons of high
molecular weight rather than on the gaseous low-molecular-weight
compounds, as it is likely that the latter will not contribute significantly
to the net flux from the air to the sea. Scientists should be encouraged
to obtain detailed spectra of the individual hydrocarbons by using
state-of-the art instrumental techniques; these data will make easier the
task of determining the origin of the hydrocarbons, especially with
regard to the distinction between biogenic and petroleum sources.

2. A greater effort must be made to characterize the areal and
vertical distribution fields of gas-phase and particulate-phase hydro-

carbons. Especially lacking are measurements in the southern hemisphere, a region where petroleum production and consumption can be expected to increase in the future. Also, short-range transport and deposition should be measured in field studies performed in coastal regions relatively close to major urban sources. A prime site for such a study is the area off the northeast coast of the United States.

3. We anticipate that, eventually, measurement programs will be carried out by a large number of individuals representing research and monitoring groups for many disciplines. It is essential that some mechanisms be established for the distribution of appropriately formulated analytical standards to these groups for the purpose of intercalibration.

4. The reaction kinetics of hydrocarbons in the atmosphere must be better understood if we are to determine the lifetime and fate of the various hydrocarbon classes. Important areas of study are gas-to-particle conversion and the distribution coefficients between solid and gas phases for various hydrocarbon species.

## REFERENCES

Bidleman, T. F., and C. E. Olney (1974). High-volume collection of atmospheric polychlorinated biphenyls, submitted to *Bull. Environ. Contam. Toxicol. 11*, 442.

Duce, R. A. (1973). Atmospheric hydrocarbons and their relation to marine pollution, *Background Papers for a Workshop on Inputs, Fates, and Effects of Petroleum in the Marine Environment*, Vol. 1, p. 416, National Academy of Sciences, Washington, D.C.

Duce, R. A., E. J. Hoffman, J. L. Fasching, and J. L. Moyers (1974a). The collection and analyses of trace elements in atmospheric particulate matter over the North Atlantic Ocean, in World Meteorological Publication 368, p. 370.

Duce, R. A., G. Quinn, and L. Wade (1974b). Residence time of non-methane hydrocarbons in the atmosphere, *Marine Pollut. Bull. 5*, 59.

Elliot, W. P., F. L. Ramzey, and R. Johnston (1974). Particle concentrations over the oceans, *J. Rech. Atmos. 8*, 939.

Eschenroeder, A. (1974). Survey of the state of knowledge of naturally emitted reactive hydrocarbons in the atmosphere, General Research Corp., Santa Barbara, Calif.

Garrett, W. D., and V. M. Smagin (1976). *Determination of the Atmospheric Contribution of Petroleum Hydrocarbons to the Oceans*, World Meteorological Organization Report No. 440, Geneva, Switzerland, 27 pp.

Hogan, A., V. Mohnen, and V. Schaefer (1973). Comments on oceanic aerosol levels deduced from measurements of electrical conductivity of the atmosphere, *J. Atmos. Sci. 30*, 1455.

Hogan, A. (1975). Aitken nuclei in the marine atmosphere, in *Background Papers for a Workshop on the Tropospheric Transport of Pollutants to the Ocean*, NRC Ocean Sciences Board, p. 288.

Hidy, G. M., and I. R. Brock (1970). An assessment of the global sources of tropospheric aerosols, 2nd Clean Air Congress, Washington, D.C., December, 36 pp.

Junge, C. (1975). Basic considerations about trace constituents in the atmosphere as related to the fate of global pollutants, paper given at the ACS meeting, Philadelphia, Pa., Apr. 1975, published in *Fate of Pollutants* (Advances in Environmental Science and Technology Series), I. H. Suffet, ed. Wiley-Interscience, New York.

Ketseridis, G., J. Hahn, R. Jaenicke, and C. Junge (1976). Organic constituents of atmospheric particulate matter, Atmos. Environ. *10*, 603–610.

Kopczynski, S. L., W. A. Lonneman, T. Winfield, and R. Seila (1975). Gaseous pollutants in St. Louis and other cities, *J. Air Pollut. Control Assoc. 25*, 251–255.

Lamontagne, R. A., J. W. Swinnerton, and V. J. Linnenbom (1974). $^1C$–$^4C$ hydrocarbon in the North and South Pacific, *Tellus 25(1–2)*, 71–77.

Lonneman, W. A., S. L. Kopczynski, P. E. Darley, and F. D. Sutterfield (1974). Hydrocarbon composition of urban air pollution, *Environ. Sci. Technol 8*, 229–236.

Machta, L. (1975). A computer program for estimating pollutant transfer from air to sea, in *Background Papers for a Workshop on the Tropospheric Transport of Pollutants to the Ocean*, NRC Ocean Sciences Board, p. 227.

McAuliffe, C. D. (1973). Partitioning of hydrocarbons between the atmosphere and natural waters, in *Background Papers for a Workshop on Inputs, Fates and Effects of Petroleum in the Marine Environment*, Vol. I, p. 280, National Academy of Sciences, Washington, D.C.

McDonald, W. F. (1938). *Atlas of Climatic Charts of the Oceans*, U.S. Dept. of Agriculture, Weather Bureau, Washington, D.C.

Ocean Affairs Board (1975). *Petroleum in the Marine Environment*, report of a Workshop on Inputs, Fates and Effects of Petroleum in the Marine Environment, National Academy of Sciences, Washington, D.C.

Prospero, J. M. (1975). Diagnostics of particulate transport, in *Background Papers for a Workshop on the Tropospheric Transport of Pollutants to the Ocean*, NRC Ocean Sciences Board, p. 243.

Peterson, I. T., and C. Junge (1971). Sources of particulate matter in the atmosphere, in *Inadvertant Climate Modification, Study of Man's Impact on the Climate*, W. H. Matthews, W. W. Kellogg, and G. D. Robinson, eds., MIT Press, Cambridge, Mass.

Rasmussen, R. A., and F. W. Went (1965). Volatile organic material of plant origin in the atmosphere, *Proc. Nat. Acad. Sci. U.S. 53*, 216.

Rasmussen, R. A. (1972). What do the hydrocarbons from trees contribute to air pollution, *J. Air. Pollut. Control. Assoc. 22*, 537.

Rasmussen, R. A. (1974). Contributions of natural organics to the global air pollution burden, Air Pollution Control Association Mtg., Denver, Colo., June 1974.

Ripperton, L. A., O. White, and H. E. Jeffries (1967). Proc. Div. Water, Air, and Waste Chem., 147th ACS Mtg., Chicago, Ill., 54 pp.

Robinson, E., and R. C. Robbins (1968). *Sources, Abundance, and Fate of Gaseous Atmospheric Pollutants*, American Petroleum Institute, New York, 123 pp.

Stephens, E. R., and F. R. Burleson (1967). Analysis of the atmosphere for light hydrocarbons, *J. Air. Pollut. Control. Assoc. 17*, 147–153.

Swinnerton, J. W., and V. J. Linnenbom (1967). Determination of $^1C$–$^4C$ hydrocarbons in sea water by gas chromatography, *J. Gas Chromatog. 5*, 570–574.

Swinnerton, J. W., and R. A. Lamontagne (1974). Oceanic distribution of low-molecular-weight hydrocarbons; baseline measurements, *Environ. Sci. Technol. 8*, 657–663.

USDHEW (1970). *Air Quality Criteria for Hydrocarbons*. U.S. Dept. of Health, Education and Welfare, Public Health Service, Environmental Health Service, Nat. Air Pollut. Control. Admin., Washington, D.C.

# 8 Nonhydrocarbon Gases

## I. INTRODUCTION

The gases to be considered are various species of sulfur, nitrogen, and carbon. At the outset, it can be stated that the nonhydrocarbon gases studied here do not directly alter the chemical properties of the oceans in any significant way. However, these gases react in the atmosphere with water, ozone, and their related free-radical species to form products that could have a profound effect on climate.

*The central species in this chapter is sulfur dioxide, which is emitted as a pollutant, mainly over continents, and which becomes oxidized to sulfuric acid and sulfate in the atmosphere. These end products form aerosols that could have a direct influence on climate by altering the radiation balance of the earth.* Also, because of the hygroscopic nature of sulfates, *these aerosols could influence climate indirectly by affecting cloud microphysical process and dynamics, thereby changing the albedo (World Meteorological Organization, 1975). All the other gases considered in this chapter have been included because they might significantly affect the rate of conversion of $SO_2$ to particulate sulfates.* For example, the oxides of nitrogen, NO and $NO_2$ (called $NO_x$ for convenience), and CO are emitted as pollutants. The $NO_x$, CO, and the natural precursors of CO ($CH_4$, $H_2CO$, and various free-radical

Members of the Working Group on Nonhydrocarbon Gases were J. P. Friend, *chairman*; D. D. Davis, C. S. Kiang, and W. Seiler.

181

species) interact with atmospheric radiation, ozone, and water vapor to produce OH (hydroxyl-free radical); OH in turn affects the oxidation rate of $SO_2$ to $H_2SO_4$ via homogeneous gas-phase reactions. Another pathway for $SO_2$ oxidation is a heterogeneous oxidation involving liquid water (as droplets), oxygen, and ammonia or trace amounts of catalytic ions such as $Fe^{3+}$ and $Mn^{2+}$. The rates of $SO_2$ conversion via the homogeneous and heterogeneous pathways are most likely of equivalent magnitude (of the order of $10^{-3}$ % $sec^{-1}$). Over the oceans, all of these substances are present in relatively constant, or slowly varying, concentration, which is a consequence of the dispersed, or large-scale, nature of the sources; over the continents, the sources are much more variable in space and time.

Man's contribution of $SO_2$ to the global flux of atmospheric gaseous sulfur compounds (which occurs mainly as $SO_2$, $H_2S$, and organic sulfides) is of the order of 35 percent (Friend, 1973) to 45 percent (Granat et al., 1976). The rate of growth in the use of fossil fuels containing sulfur is such that anthropogenic $SO_2$ emissions will soon dominate the global cycle of sulfur through the atmosphere. In addition, the increased use of tall stacks to aid in the dispersion of $SO_2$ emissions (so that local pollution control regulations are not violated) will increase the quantity of $SO_2$ and sulfate particulate matter injected above the planetary boundary layer; this will, in effect, increase the efficiency with which these materials are transported to the ocean environment.

Another factor of possible importance is that the increasing $SO_2$ emissions may consume ammonia, via the heterogeneous oxidation process, at such a rate as to cause significant reduction in $NH_3$ concentrations in the atmosphere. This would serve to increase the residence time of $SO_2$ in the atmosphere and, thus, increase transport to the ocean environment.

In the global cycles of the sulfur, nitrogen, and carbon compounds, the oceans can act either as a source or a sink (or both). However, these cycles are, for the most part, poorly understood. Consequently, an assessment of the global impact of pollutants cannot be made without a thorough study of the fluxes of these substances into, and out of, the oceans and of the concentrations of these and related substances in the atmosphere and in surface ocean waters.

Carbon dioxide is an important nonhydrocarbon constituent of the atmosphere that has been the object of considerable concern and attention because of the potential impact of increasing anthropogenic emissions on global climate. The oceans play an especially important role in regulating the concentration of $CO_2$ in the atmosphere. It is generally assumed that the effects of the projected increase in atmospheric $CO_2$ on the oceans will be small and within the range of tolerability for marine biota (MacIntyre, 1970). However, Sundquist and Southam (1976) suggest that the pH of the oceans could decrease by as much as 1 pH unit over the next 100 years. Because of extensive discussion elsewhere, $CO_2$ will not be considered here. The interested reader is referred to discussions by Keeling and by Schneider and Kellogg [see Rasool (1973)] and to a treatment of the subject by Machta (1971). A good recent treatment of the possible effects of $CO_2$ increases on climate is given by Manabe and Wetherald (1975).

The panel did not discuss nitrous oxide, $N_2O$, in any detail during its deliberations. A brief discussion of atmospheric concentrations of this naturally occurring trace gas appears in Section II. It is worth noting that recently there has been concern over the generation of excess $N_2O$ by soil microorganisms (the natural source) because of the use of artificial fertilizers [cf. Crutzen (1974); McElroy *et al.* (1975); see also Svensson and Soderlund (1976) for a discussion of the global cycle of nitrogen]. The basis of the concern is that $N_2O$ is the major precursor to oxides of nitrogen, which act, in the stratosphere, as catalysts for reactions that lead to the destruction of stratospheric ozone. Thus, continued use of fertilizer might appreciably decrease the concentration of stratospheric ozone. Controversy has arisen over several aspects of $N_2O$ production and destruction or removal. [In addition, see Crutzen (1976); Liu *et al.* (1976); Sze and Rice (1976).] One of the most important areas of concern and controversy is the question of whether the oceans act as a major source of $N_2O$ [as Hahn (1974) suggests] or if they act, in net, as a major sink of $N_2O$ [as maintained by McElroy *et al.* (1975)]. The oceanic source of $N_2O$, whether major or minor, is, like the continental source, thought to be due mainly to nitrate-reducing microorganisms. However, Hahn and Junge (1977) suggest that nitrification by microorganisms might also produce significant amounts of $N_2O$. While there is little general agreement about how the $N_2O$ cycle works, there is a universal consensus that a much better understanding of the natural phenomena operative within the cycle is needed before a rational assessment can be made of the environmental effects of fertilizer-produced $N_2O$.

## II. SPECIES OF INTEREST

The background concentrations of several of the species of interest in tropospheric air are summarized in Table 8.1.

TABLE 8.1 Background Concentrations of Various Tropospheric Gases (Units: parts per billion, volume)

| Species | Temperate Zone | Tropical Zone | Polar Zone |
|---|---|---|---|
| $H_2S$ | 0.002–0.20 (1), (2), (10) | 0.006 (1) | — |
| $SO_2$ | 0.2–5.0 (2), (5), (8) | 0.1–2.0 (3), (6), (9) | 0.002–0.02 (4) |
| $N_2O$[a] | 200–400 (14) | 250 (13) | 240–260 (13) |
| $NO_2$ | <1–3.0 (2), (7), (8) | <0.5–5.0 (3), (6), (9) | <0.2–2.0 |
| $NH_3$ | <1–10 (2), (8) | 1–>20 (3), (6), (9) | — |
| $DMS$[b] | 0.03–1 (10) | — | — |
| $OCS$ | 0.03–0.1 tentative (10) | — | — |
| $CS_2$ | 0.03–0.3 (10) | — | — |
| $CO$ | 60–80 southern hemisphere (11) | 70–120 (11) | <60 (South Pole) estimated from (11) |
| $O_3$[c] | 30 (20–40)[d] | 15 | 25 |
| (12) | | | |
| $OH$[e] | $1.3 \times 10^6$ molecules/cm$^3$ (12) | $3 \times 10^6$ molecules/cm$^3$ (12) | $<5 \times 10^5$ molecules/cm$^3$ (12) |

[a] Average results only reported for tropical maritime air.
[b] DMS = dimethylsulfide, $(CH_3)_2S$.
[c] Ozone displays variability in many scales, most notably diurnally and seasonally. The concentrations shown in the table are typical values for near-surface air.
[d] Temperate latitudes show the largest seasonal variation, and the approximate range of typical values is shown. The concentrations are highest in summer and lowest in winter.
[e] Diurnally and seasonally averaged 3-km values from the 2-D OH model of Chang and Wuebbles (1976).

(1) Natusch et al. (1972); Slatt et al. (1977).
(2) Breeding et al. (1973).
(3) Lodge and Pate (1966).
(4) Ba Cuong et al. (1974).
(5) Georgii (1970).
(6) Pate et al. (1970).
(7) Ripperton et al. (1968).
(8) Junge (1956).
(9) Junge (1960).
(10) Maroulis et al. (1977).
(11) Seiler (1976).
(12) Chang and Wuebbles (1976).
(13) Hahn (1974).
(14) Schutz et al. (1970).

## A. SULFUR COMPOUNDS

### 1. Sulfur Dioxide, $SO_2$

Sulfur dioxide is injected into the atmosphere as a pollutant, mainly as a consequence of the burning of sulfur-bearing fossil fuels, the refining of petroleum, and the smelting of sulfide ores. Volcanoes and geothermal vents also emit $SO_2$ but in considerably smaller amounts than those from pollution sources. The reduced sulfur gases of biogenic origin (discussed below) may, in part, become oxidized to $SO_2$ in the atmosphere. As of 1970, the source strengths for these various processes are estimated to be $65 \times 10^{12}$ g yr$^{-1}$ from pollution sources and $2 \times 10^{12}$ g yr$^{-1}$ from volcanoes (Friend, 1973). The amounts of $SO_2$ produced by the oxidation of biogenically produced compounds is difficult to estimate because of uncertain and unknown factors governing source strengths and reaction mechanisms in the atmosphere. It has been commonly assumed, in studies of global cycles, that the reduced sulfur compounds are ultimately oxidized to sulfate in the atmosphere; however, the question of how much intermediate $SO_2$ may be created in the process has been ignored.

Concentrations of $SO_2$ over the Atlantic Ocean were found to range up to 3 parts per billion mass (ppbm $= 10^{-9}$ g of $SO_2$ per g air), or about 3.7 $\mu$g m$^{-3}$. The concentration reaches a maximum at about 50° N latitude and diminishes to values approaching the detection limit northward and southward of this latitude (Matthews *et al.*, 1971). Table 8.2 shows concentrations of $SO_2$ and $SO_4^{2-}$ reported by Ba Cuong *et al.* (1974) for various marine atmospheres. Lodge *et al.* (1974) also reported concentrations of about 0.05 part per billion, volume (ppbv $= 10^{-9}$ m$^3$ of $SO_2$ per m$^3$ air STP) for Panama.

*2. Hydrogen Sulfide, $H_2S$*

$H_2S$ can be produced in major quantities in reducing sedimentary environments; however, there is considerable debate as to whether $H_2S$ can escape in significant quantities through the overlying oxygen-rich surface waters. Brinkmann (1974) has shown that $H_2S$ can be emitted from a tropical jungle lake during periods when the water level

TABLE 8.2 Atmospheric Concentrations of $SO_2$ and Sulfate after Ba Cuong *et al.* (1974) (Units: $10^{-6}$ g m$^{-3}$ air, STP)

| Location | $SO_2$ | $SO_4^{2-}$ |
|---|---|---|
| Antarctic (60–70° S) | 0.13 | 1.57 |
| Subantarctic (40–60° S) | 0.18 | 1.68 |
| South Pacific (20–40° S) | 0.12 | 1.15 |
| North Atlantic (50–8° N) | — | 2.33 |
| Mediterranean Sea | 2.27 | 8.43 |

is low because of evaporation. However, $H_2S$, although present in high concentration in bottom waters of the Baltic and Black Seas, has not been detected in the surface waters or in the atmosphere above (Fonselius, 1972; see also Ostlund and Alexander, 1963). Fonselius (1972) also concludes that the amounts of $H_2S$ in stagnant ocean basins such as the Black Sea are much too small to account for the amount of sulfur seemingly required by the global budget.

It has been generally accepted that $H_2S$ is emitted by organisms into the atmosphere from swamps, marshes, and muds associated with intertidal flats. However, reliable measurements of $H_2S$ concentrations in the atmosphere are virtually nonexistent. Bandy (1977) recently measured mean concentrations of 58 pptv off Wallops Island over a period of a few days. Slatt *et al.* (1977) found concentrations of $H_2S$ averaging about 6 pptv at Barbados and at Sal Island; at a coastal site in Miami and over the Everglades, concentrations averaging 46 pptv were found.

*3. Organic Sulfides*

The compounds methyl mercaptan, $CH_3SH$, and dimethyl sulfide (DMS), $(CH_3)_2S$, have been found to be produced by terrestrial and aquatic biological systems (cf. Lovelock *et al.*, 1972). There apparently are several possible biological and biochemical processes that can produce organic sulfide gases. The low solubilities of these compounds in water would favor their evolution from the marine environment. However, their oxidation reactions in water have not been studied, and losses as a result of such reactions cannot be estimated. The photooxidation of DMS in air was studied by Cox and Sandalls (1974), who could not find $SO_2$ as a product. This suggests that oxidation of DMS in the atmosphere may lead to the formation of sulfinic and sulfonic compounds rather than $SO_2$. The behavior of methyl mercaptan under photooxidation has not been studied.

The only measurements of DMS in the open atmosphere that are known to the panel are those of Maroulis and Bandy (1977), who found concentrations in the range of 0.05–0.3 ppbv (mean, 0.058) off Wallops Island, Virginia.

The strength of the global sources of biogenic sulfur gases ($H_2S$ and organic sulfides) has been estimated by Friend (1973) to be about $110 \times 10^{12}$ g of sulfur per year from both the land and the oceans, $60 \times 10^{12}$ and $50 \times 10^{12}$ g yr$^{-1}$ from each source, respectively. On the other hand, Granat *et al.* (1976) estimated the biogenic sulfur flux to be considerably less—$32 \times 10^{12}$ g yr$^{-1}$, of which $5 \times 10^{12}$ g is from land and

$27 \times 10^{12}$ g is from the ocean. These source estimates were derived indirectly; they are simply the fluxes required to make the world budget balance. However, many of the source strengths used in the budget are based on a very few isolated data and require further substantiation. Consequently, the land and ocean source strength estimates are, themselves, open to considerable question.

B. NITROGEN CÖMPOUNDS

*1. Nitric Oxide, NO*

Natural concentrations appear to be in the range 50–200 pptv (Noxon, 1975), much lower than earlier estimated (i.e., 1 ppbv). The concentration of NO can be expected to be extremely variable because of the reactivity of NO toward $O_3$ and the presence of strong anthropogenic sources.

*2. Nitrogen Dioxide, NO$_2$*

The discussion for NO applies also to $NO_2$ because the two are related photochemically via the reactions:

$$NO + O_3 \rightarrow NO_2 + O_2,$$
$$NO_2 + h \rightarrow NO + O,$$
$$O + O_2 \rightarrow O_3.$$

*3. Nitric Acid, HNO$_3$*

To the best knowledge of the panel members, no data exist for the natural troposphere. Concentrations in the troposphere, according to recent models, could range from $10^8$ to $10^{10}$ molecules cm$^{-3}$.

*4. Ammonia, NH$_3$*

The principal sources of atmospheric ammonia have been thought to be combustion and the decay of biological matter in the soil. Estimates of the land source strength range from $30 \times 10^{12}$ to $200 \times 10^{12}$ g yr$^{-1}$ of nitrogen [see Hahn and Junge (1977) for a summary of sources of $NH_3$ and other nitrogen gases]. Robinson and Robbins (1970) estimated that about $3.2 \times 10^{12}$ g yr$^{-1}$ of $NH_3$ nitrogen comes from combustion sources. Healy *et al.* (1970) suggested that the $NH_3$ concentrations measured in the United Kingdom could be largely accounted for by the decay of urea in animal urine deposited on soil. Because most of the urine comes from cattle and other animals that are raised for man's consumption or pleasure, these emissions probably should be classified as "anthropogenic," as was done by Tsunogai (1971).

The concentrations of $NH_3$ over the oceans have been measured by Tsunogai (1971) and found to vary from about 0.4 to 6 ppb. The higher concentrations in the range were attributed to continental influences. The low concentrations may, in fact, represent an equilibrium of $NH_3$ with oceanic $HN_4^+$ and metallic–ammonia ionic complexes, but this is quite uncertain.

Other measurements of background $NH_3$ concentrations are summarized in Table 8.1.

There are many aspects of the atmospheric cycle of $NH_3$ that are poorly characterized: the total fluxes, the distribution of sources, the regulatory effect of equilibria over the oceans, the uptake by soil and vegetation, the deposition rates in precipitation. However, if the picture given by Hahn and Junge is substantially correct, then the $NH_3$ flux comprises a significant component of the global atmospheric nitrogen cycle.

### 5.  Nitrous Oxide, $N_2O$

Surface air measurements of $N_2O$ have been summarized by Schutz et al. (1970), by Hahn (1974), and, most recently, by Hahn and Junge (1977). Most of the measurements have been made in temperate latitudes in continental air, and they exhibit some degree of variability, ranging over a factor of 2 in concentration, with an average value of close to 260 ppb. The few measurements available from the maritime tropics and polar latitudes yield $N_2O$ concentrations that average close to 250 ppb and have very little scatter about the means compared with the measurements made in temperate latitude continental air. Recent measurements by Rasmussen and Pierotti (1977) also yield maritime air concentrations with low variability but with a mean of 330 ppb. The differences between Rassmussen's results and those of other investigators appear to be attributable to differences in analytical procedures and to the lack of intercalibration.

### C.  OTHER GASES

### 1.  Carbon Monoxide, CO

The concentrations, distribution, sources, and sinks of CO are reviewed in detail by Seiler (1976). The main points are summarized here.

*Concentration and Distribution* During the last five years, atmospheric CO measurements were made during the couse of several expeditions, using ships, aircraft, and balloons. Figure 8.1 shows the mean latitudinal distribution of CO over the northern and southern Atlantic. Only those CO data were used that were obtained in marine air masses. The latitudinal distribution shows a maximum at 50° N latitude at the surface. The CO mixing ratio decreases slightly with altitude to about 0.12–0.15 ppmv in the upper troposphere at 50° N; the upper troposphere values in the South Atlantic are not very different from the surface values, about 0.06–0.07 ppmv at 60° S. Recent measurements over the Pacific Ocean between 60° N and 60° S indicated a similar distribution. Consequently, Figure 8.1 can be considered to be representative of global conditions. Above the tropopause, the CO mixing ratio decreases to about 0.04–0.05 ppmv and remains constant up to altitudes of 5 km above the tropopause (Seiler and Warneck, 1972). In contrast to the situation in the troposphere, there seems to be a uniform distribution of CO in the lower stratosphere in the two hemispheres. Large fluctuations of the CO mixing ratio in air were found only in the lower atmosphere of the northern hemisphere. Here, the

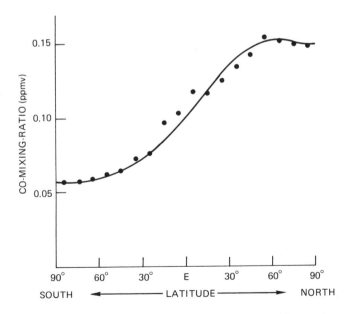

FIGURE 8.1 Latitudinal distribution of carbon monoxide over the Atlantic Ocean (Seiler, 1974).

TABLE 8.3 Estimated Source Strengths for Atmospheric Carbon Monoxide

| Source | Strength ($10^{14}$ g yr$^{-1}$) | Reference |
|---|---|---|
| Pollution | 6 | Seiler and Zankl (1976) |
| Marine microorganisms | 0.2–2 | Seiler and Schmidt (1974) |
| Fires (forests, brush, agricultural waste) | 0.6 | Robinson and Robbins (1967) |
| Oxidation of hydrocarbons | 0.6 | Robinson and Robbins (1967) |
| Direct emission from higher plants | 0.2–2 | Seiler and Giehl (1976) |
| | | Fischer and Seiler (1975) |
| Oxidation of methane: | | |
| Using (OH) = $2.5 \times 10^6$ cm$^{-3}$ | 22 (northern hemisphere) | Weinstock and Chang (1974) |
| Using (OH) = $(1–3) \times 10^6$ cm$^{-3}$ | 15–40 | Levy (1971); McConnell et al. (1971) |
| Using (OH) = ? | 4 | Seiler (1976) |

CO mixing ratio varies by a factor of 2 within a few hours, even over the Central Atlantic. These fluctuations are most likely attributable to the transport of polluted air parcels over the ocean areas. Although the patchiness of the distribution of CO-producing marine organisms could contribute to the variability, it seems unlikely that there would be major differences between the two hemispheres in this regard.

*Sources*    The sources of CO have been estimated by various workers, and their findings are summarized and reviewed by Seiler (1975). The major global sources of CO are generally agreed to be pollution (primarily from automobiles) and oxidation of atmospheric methane by OH radicals. Lesser, but not negligible, sources are forest fires and marine microorganisms. Siphonophores (jellyfish), suggested as major producers of marine CO, have been found to be quite negligible sources (Seiler, 1975). Terrestrial plants emit hydrocarbons that are oxidized to CO in the atmosphere; some plants may be capable of producing CO directly. This is discussed further in Section V.

There is no universal agreement among the various investigators as to the relative importance of pollution sources. The lack of agreement is attributable to the widely differing estimates of the quantity of CO that is produced by methane oxidation. The various estimated source strengths are given in Table 8.3. If the estimates by Levy (1971), McConnell *et al.* (1971), and Weinstock and Chang (1974) are correct, then pollution sources constitute less than a third of the total. If, on the other hand, Seiler (1975) is correct, then the pollution sources represent about half of all CO emissions. The mean residence time of CO corresponding to Seiler's source estimates is about 0.4 year using a mean CO mixing ratio of 0.11 ppm for the troposphere. The contention that pollution sources supply a large fraction of the CO budget is supported by CO mixing ratios over the continents, which, even in rural areas, are sometimes factors of 2 to 5 higher than over the oceanic regions in the same latitudes. It is also consistent with the apparent dominance of CO sources in the northern hemisphere relative to the southern hemisphere as evidenced in the meridional profile of Figure 8.1. Because of the relative land mass distributions, methane emissions are expected to be somewhat larger in the northern than in the southern hemisphere. However, the differences are not thought to be large enough to account for the observed latitudinal gradient in CO. Because of the difference in concentrations between the hemispheres and between continental and marine air, and because of the relatively long residence time of CO in the atmosphere, it would be an ideal tracer for studying global transport.

## 2. Peroxide, $H_2O_2$

To the best knowledge of the panel members, no measurements of $H_2O_2$ in the natural troposphere have been reported. Based on model predictions (cf. Levy, 1971), concentration values ranging from $10^7$ to $10^{10}$ molecules/cm$^3$ could be expected.

## 3. Ozone, $O_3$

The concentration of $O_3$ is quite variable throughout the troposphere, a consequence of the variability of its photochemistry and the anthropogenic sources of hydrocarbons. Natural levels of ozone are believed to be in the 25 to 30 ppbv range, on the average, with a principal source being the stratosphere. Values 10 to 20 times larger than this natural level are now being observed in many industralized countries because of the complex areal distribution and temporal variability of $NO_x$ and hydrocarbon sources.

## 4. Hydroxide Ion, OH

Natural tropospheric measurements of the OH free radical have only become available within the last year (Davis et al., 1976). These measurements are limited to a latitude range of 31 to 21° N and altitudes of 7 to 11.5 km. Values ranged from $3 \times 10^6$ to $8 \times 10^6$ radicals cm$^{-3}$ in October 1975. It is to be expected that the OH concentrations will be quite variable. However, recent budget considerations of $CHCl_3$ imply much smaller average OH concentrations: $3 \times 10^5$ per cm$^3$ in the northern hemisphere and $7 \times 10^5$ per cm$^3$ in the southern hemisphere (Crutzen and Fishman, 1977). Substantially higher OH concentrations can only be reconciled with our present knowledge of the CO budget by the introduction and addition of major sources of CO.

## III. TRANSPORT

The lifetimes of many important species are sufficiently long that regions remote from the sources could be affected by the species themselves or by their reaction products. For those species whose lifetimes are shorter than about a few weeks, significant concentration differences can exist over horizontal scales smaller than the dimensions of continents or oceans. The dynamical aspects of transport and transport modeling are discussed in detail in Chapter 3. *The Panel emphasizes that such models, if they are to be realistic, must take into account chemical reactions that occur in the atmosphere as a consequence of interactions of reactive "plume" components with species distributed throughout the ambient atmosphere and also with radiation and clouds. The representation of such reaction systems in models is complicated and, perhaps, for some species is not amenable to shortcut parameterizations.*

## IV. TRANSFORMATION

There are several processes by which sulfur dioxide can be transformed in the atmosphere into sulfuric acid and various sulfates. Among these are gas-phase bimolecular and termolecular reactions, catalytic and noncatalytic oxidation in droplets, and catalytic oxidation of the gas–surface interface of particles. The factors that influence the rate of sulfate production are gas-phase transports, gas-phase reaction cross sections, liquid-phase transports, reaction rates in the liquid phase, nucleation, and surface catalysis. It is also necessary to consider the manner in which each mechanism is affected by atmospheric conditions. With the appropriate mathematical formulations, computations can be made of the contribution of each mechanism to the overall process. In this fashion, one can hope to gain some insights about these complicated systems.

A schematic block diagram for the chemical and physical transformation of SO$_2$ to sulfates (at relative humidity less than 100 percent) is shown in Figure 8.2. In the diagram, the arrows represent the physical or chemical processes and the boxes represent the products in various phases. In the following discussion, the chemical transformations are separated into two categories: homogeneous gas-phase processes and heterogeneous processes.

### A. CHEMICAL TRANSFORMATIONS

#### 1. Homogeneous Gas-Phase Processes

As outlined in Section I, the long-range global impact of the nonhydrocarbon pollutants appears to reside in their modulating effect on tropospheric aerosols, specifically their mass concentration, composition, number density, and distribution in the horizontal and the vertical. The purpose of the following section is to outline briefly the role that homogeneous gas-phase chemical processes play in affecting these aerosol characteristics.

The principal process of concern here is the homogeneous gas-phase

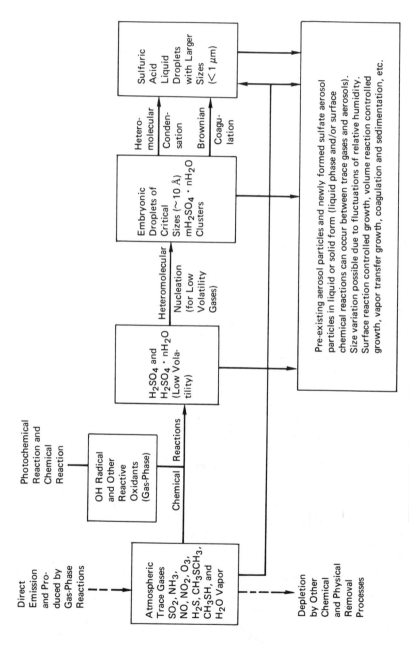

FIGURE 8.2 Gas-to-particle conversion (chemical and physical transformation).

194

conversion of $SO_2$ to $H_2SO_4$ molecules. This question has been extensively explored by Davis and Klauber (1975), with the conclusion that the single most important gas-phase reaction process is

$$OH + SO_2 + M \rightarrow HSO_3 + M. \qquad (8.1)$$

The detailed chemistry that follows Reaction (8.1) is still somewhat speculative (Davis and Klauber, 1975), but it is commonly accepted that $H_2SO_4$ molecules are formed rapidly after Reaction (8.1) has taken place. Thus, the major focal point of the following discussion will be how various parameters affect the rate of process (8.1). In the most simplistic sense, this rate can be expressed as

$$-\frac{d(SO_2)}{dt} = \frac{d(H_2SO_4)}{dt} = k(OH)(SO_2)(M),$$

where $(M) = N_2$ and the three key parameters are the bimolecular rate constant, $k$, the steady-state concentrations of the hydroxyl free radical, and the absolute concentration of $SO_2$. The values of $k$ in this case have been measured at a temperature of 300 K (Davis and Klauber, 1975). Gaseous $SO_2$ concentrations in the natural troposphere (see Tables 8.1 and 8.2) are not so well known as would be required for reliable calculations of the global $H_2SO_4$ production. By far the least-known quantity in Reaction (8.1) is the steady-state concentration of OH. This free radical is highly reactive with several atmospheric trace gases and has an average lifetime of approximately 1 sec. Because ozone, water vapor, and uv radiation are required to produce OH (see Figure 8.3), we can expect the OH concentration to depend strongly on altitude, latitude, and solar angle. It is readily apparent that direct measurements of the OH radical would be of great importance in improving the calculations of the gas phase $SO_2$-to-$H_2SO_4$ conversion processes. At present, only one set of measurements has been reported for the natural troposphere (Davis *et al.*, 1976).

As can be seen from Figure 8.3, a large number of atmospheric gases also act, directly or indirectly, to modulate the OH steady-state concentration and, hence, modulate the rate of conversion of $SO_2$ to $H_2SO_4$. Most important are those gases involved in the production and removal of OH. The major atmospheric source is believed to be the photolysis of $O_3$ at wavelengths less than 3100 Å. At these wavelengths, $O(^1D)$ is produced, a small fraction of which reacts with $H_2O$ to generate two OH radicals. [It should be noted that this source term is critically dependent on the ratio of $H_2O$ to $N_2$ in the atmosphere because of the deactivation of $O(^1D)$ by $N_2$, which gives the less

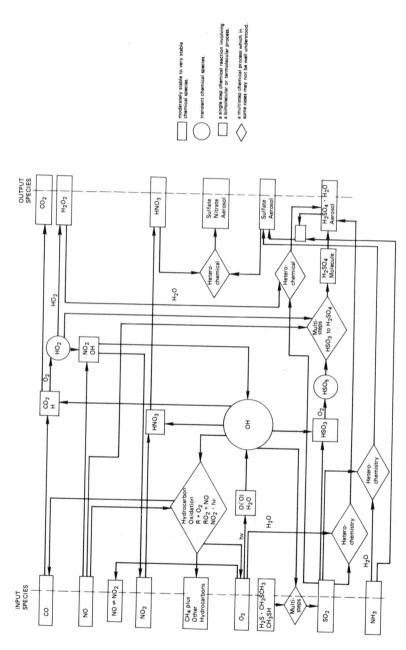

FIGURE 8.3 Chemical transformations of gases to aerosol involving OH radicals (from work by D. D. Davis and C. S. King).

196

reactive form of atomic oxygen, $O(^3P)$.] From the above discussion, it is apparent that anything that affects tropospheric $O_3$ will also have a significant effect on the OH concentrations. Figure 8.3 indicates that hydrocarbons (natural and anthropogenic) and $NO_x$ species will have a significant impact on tropospheric $O_3$ concentrations. The most important removal processes would now appear to be those involving the reaction of OH with CO and $CH_4$:

$$OH + CO \rightarrow CO_2 + H, \qquad (8.2a)$$

$$OH + CH_4 \rightarrow H_2O + CH_3. \qquad (8.2b)$$

However, whether or not CO is an efficient sink for OH depends largely on the fate of the $HO_2$ radical formed from the reaction of H with $O_2$:

$$H + O_2 \overset{M}{\rightarrow} HO_2. \qquad (8.3)$$

It was once thought that the principal reaction was

$$HO_2 + NO \rightarrow NO_2 + OH, \qquad (8.4)$$

which would simply regenerate the OH radical and, hence, would argue against Reaction (8.2a) being an important modulator of the OH concentration. This line of thought now has to be re-examined in view of the recent measurements by Noxon (1975), who reported natural tropospheric concentrations for $NO_2$ that were 10 to 30 times less than previously believed. Sze (1977) addresses this same point and suggests that with increasing CO, less $CH_4$ and $CH_3Cl$ would be destroyed in the troposphere, thereby ultimately causing increased $O_3$ destruction in the stratosphere through the enhancement of various photochemical cycles. Should this generally be the case, one would expect the $HO_2$ radical to undergo chain termination via the processes:

$$HO_2 + HSO_5 \rightarrow H_2SO_5 + O_2, \qquad (8.5)$$

$$HO_2 + HSO_4 \rightarrow H_2SO_4, \qquad (8.6)$$

$$HO_2 + HO_2 \rightarrow H_2O_2 + O_2. \qquad (8.7)$$

The above discussion shows the importance of both CO and $NO_x$ in the OH cycle and, thereby, the gas-phase $SO_2$ conversion process (8.1). Although a more detailed discussion of other chemical species will not be given here, *it is apparent from Figure 8.3 that long-range predictions of sulfate aerosol formation could be largely dependent on long-term changes in the concentrations of many other pollutant gases.* This could be especially true for CO, where already the northern hemisphere concentrations are nearly a factor of 3 greater than those of the southern hemisphere (Seiler, 1975).

## 2.  Heterogeneous Processes

The term "heterogeneous" reaction is used to describe chemical reactions occurring in two or more phases. Among all the transformations, heterogeneous reactions pose the most difficult problems. At present, only crude estimates can be made of the conversion rate of $SO_2$ to sulfate and the removal rate of $SO_2$ by heterogeneous reactions.

The heterogeneous reactions can be separated into two categories: gas–liquid and gas–solid surface reactions.

*Gas–Liquid Reactions*    Sulfur dioxide conversion to sulfate in pure water is a slow process, but it is known to be accelerated in the presence of cations such as ammonium, ferric, and manganous ions (e.g., see Junge and Ryan, 1958). The conversion of $SO_2$ to $(NH_4)_2SO_4$ in clouds is a typical example of this process. Although there have been experimental measurements of the overall rates of $SO_2$ oxidation in various solutions, no confirmed mechanism has been developed in which the elementary reaction steps can be shown in detail. Thus, there are no known rate constants relevant to heterogeneous oxidation of $SO_2$. Most of the present theories have treated this conversion process under equilibrium conditions and assumed the rate-determining step to be sulfite oxidation in the liquid phase. A kinetic model is needed to estimate the conversion rates under various atmospheric conditions.

*Gas–Solid Surface Reactions*    The conversion of $SO_2$ to sulfate via gas–solid surface reactions may involve physical adsorption, chemical adsorption, and heterogeneous catalysis. Rate constants for these transformation processes have not been determined. Governing parameters for determining the rate constants are the concentration of $SO_2$, the chemical nature of the gaseous medium, the chemical composition of the aerosol particles, the surface structure, the number concentration and size distribution of the particles (and, hence, their surface area). Because there is little or no information on rate constants, reliable estimates of the relative importance of heterogeneous reactions in the overall ($SO_2$–sulfate) transformation process cannot now be made.

Urgently needed are experimental measurements to identify the mechanisms and rate constants for conversion processes that may be initiated by these gas–solid surface heterogeneous reactions. It is important to emphasize here that one cannot correctly estimate the overall transformation rate of $SO_2$ to particulate sulfate if one totally neglects solid–gas heterogeneous reactions.

### B. PHYSICAL TRANSFORMATIONS

In this subsection we discuss briefly the important physical processes accompanying the oxidation of sulfur compounds and the incorporation of the products into aerosols. These processes are nucleation, growth, and coagulation. They govern the time dependence of the aerosol size distribution. (The interaction between sulfate aerosols and hydrometeors are discussed in Chapter 4 and will not be discussed here.)

### 1. Nucleation

The equilibrium vapor pressure of pure $SO_2$ is about 4 atm at room temperature. With concentrations in the atmosphere ranging from a few ppb to about 1 ppm, one would not expect $SO_2$ vapor to nucleate and form new particles in the atmosphere. However, as a result of the oxidation of $SO_2$, $H_2SO_4$ vapor can be formed. With the equilibrium vapor pressure of $H_2SO_4$ and its hydrates in the range of $10^{-6}$ Torr and less, a new particle can be formed via what has been termed heteromolecular nucleation (Kiang and Stauffer, 1973). The particles consist of molecular clusters built up from sulfuric acid and water molecules. This particle-forming process is enhanced by the presence of an ion, atom, or foreign neutral molecule of different chemical composition from that of the bulk nucleating phase.

At present, the theoretical basis of nucleation is in a rudimentary state. While homogeneous phenomena have received considerable experimental and theoretical attention, much less is known about the heteromolecular process, which is more relevant to ambient conditions. Further experimental and theoretical work in this area is urgently needed. Such work should include the obtaining of more accurate data for the equilibrium vapor pressure of pure $H_2SO_4$ and the surface tension of the solutions as a function of composition and temperature, all of which are needed for heteromolecular nucleation calculations (Kiang and Stauffer, 1973).

The possibility that other multicomponent systems may be involved in the heteromolecular nucleation also needs investigation.

### 2. Growth

Once a nucleus is formed, it can grow in mass by a heterogeneous condensation process that converts gases to particles. The mechanisms can be classified into two types: one for which growth is controlled by diffusion and one for which it is controlled by chemical reaction and diffusion.

In the first type of growth, reaction occurs in the gas phase and leads to the formation of low-volatility products that subsequently diffuse to the particle surface. For diffusion-controlled growth involving $H_2SO_4$ vapors, the above-mentioned heteromolecular effect must be considered. The important factors in applying theory for modeling purposes are the chemical state of the aerosol (i.e., its chemical composition) and the gas-phase concentrations of the various condensing species (e.g., the $H_2SO_4$ and $H_2O$ partial pressures). Another important factor is the efficiency with which the growing particle surface accommodates the impinging gas molecules. Also, equilibrium partial pressures of the low-volatility products of the relevant gas-phase chemical reactions are needed in order to apply existing theory to atmospheric problems.

In the second type of growth, gases are first adsorbed on the particle–droplet surface and subsequently react at the surface of the particle or in the interior of the droplet. Growth is controlled by surface- or volume-related reaction rates. It is difficult, if not impossible, to treat this case theoretically. Instead, experimental studies must be made to determine the necessary parameters.

### 3. Coagulation

The growth of particles by coagulation has been extensively studied both theoretically and experimentally (cf., Fuchs, 1964). Existing theories are adequate for purposes of modeling. However, if the process of coagulation is coupled with growth processes, the size-dependent composition of the aerosol becomes an important parameter. Numerical techniques for treating this type of problem are needed.

## V. REMOVAL PROCESSES

The removal of gases from the atmosphere can be effected by various chemical and physical removal processes occurring in the atmosphere and at the earth's surface. Although many chemical and physical processes can occur simultaneously for an individual gas, environmental conditions may cause one (or only a few) to be dominant. Reactive gases are primarily removed by chemical reactions forming new species in the atmosphere. Physical removal processes become important for more inert gases and for those with a high solubility in water. Among the important removal mechanisms are those heterogeneous and homogeneous reactions of gases that result in the production of aerosol particles, which are then removed from the atmosphere by physical processes. In this section, we divide the physical removal processes into two parts: wet removal processes and dry removal processes. In each case, the present knowledge of the individual removal processes is discussed. The removal rates of the individual gases are given in absolute values (g yr$^{-1}$) or in terms of the lifetime.

We adopt a global approach in which the various processes are considered as components of a worldwide budget or balance sheet. In this context, estimates of removal over land enable us to establish limits on the fluxes of substances to the oceans. The subject of removal processes is discussed in detail in Chapter 4. In this section we discuss only those aspects of the problem that relate to nonhydrocarbon gases.

### A. WET DEPOSITION

Wet removal processes are important factors in the cycle of reactive gases such as $SO_2$, $NH_3$, and $O_3$ and of gases with high solubility coefficients in water such as $NO_2$ and $H_2O_2$. For nonreactive gases with low solubility coefficients, such as CO, the deposition by wet removal is negligible. It is very efficient for those gases ($SO_2$, $NO_2$, $NH_3$) that are transformed after dissolution into less volatile species (e.g., sulfate) by heterogeneous reactions. These processes, as discussed in detail above, are also of major importance, even in nonprecipitating clouds. Wet removal processes are responsible for a total flux of $140 \times 10^{12}$ g yr$^{-1}$ of sulfur as $SO_2$ reaction products (Friend, 1973); for this flux, Granat *et al.* (1976) obtain a value of $109 \times 10^{12}$ g yr$^{-1}$.

### B. DRY DEPOSITION

Aerosol particles are removed directly from the atmosphere to the earth's surface by dry deposition, which encompasses all mechanisms other than wet removal. The dry deposition rates are partly dependent on the size of the aerosol particle and, consequently, on the settling velocity. They also depend partially on transport mechanisms in the atmosphere—turbulent transfer followed by impaction and diffusional deposition upon the surface. Dry fallout represents an important removal process for sulfate particles that are formed by physical and chemical oxidation of $SO_2$ (see above) in the atmosphere. The total removal rate of $SO_4^{2-}$ particles by dry removal is estimated to be $34 \times 10^{12}$ g of S yr$^{-1}$, which is approximately 20 percent of the total $SO_4^{2-}$ deposition from the atmosphere. Of this quantity, about $20 \times 10^{12}$ g of S yr$^{-1}$ is deposited over land and $17 \times 10^{12}$ g of S yr$^{-1}$ over the oceans (Friend, 1973); Granat et al. (1976) obtain $28 \times 10^{12}$ and $10 \times 10^{12}$ g of S yr$^{-1}$ for these areas, respectively.

The dry deposition of gases from the atmosphere includes several processes such as uptake by plants, absorption by soil microorganisms, gas exchange through the air–sea interface, and the gas exchange between troposphere and stratosphere at the tropopause. These processes are discussed in the following subsections.

## C.  UPTAKE BY PLANTS

Our knowledge about the influence of plants on atmospheric gases is very unsatisfactory. This is mainly because of the lack of appropriate data. Past experiments were carried out with such high concentrations of gas that the results cannot be considered to be representative of atmospheric conditions. In many cases, the results obtained by different authors are contradictory. In spite of these uncertainties, these experiments show that plants can affect the concentration of several gases, e.g., $SO_2$, $NH_3$, and CO. $SO_2$ and $NH_3$ are strongly absorbed by the plant material. As an example, $SO_2$ uptake by plants is estimated to be $30 \times 10^{12}$ g yr$^{-1}$ (Friend, 1973). Knowledge of the distribution of atmospheric $NH_3$ concentrations is so sparse that meaningful estimates of its uptake by vegetation and soil cannot be made (see discussion of $NH_3$ above). In support of the idea that $NH_3$ is strongly absorbed by plants, the experimental study of Hutchinson et al. (1972) led the authors to suggest that plant canopies could absorb as much as 20 kg hectare$^{-1}$ yr$^{-1}$. Absorbed CO was shown to be incorporated into the sucrose and protein of plants (Bidwell and Frazer, 1972). The uptake by plants was found to be roughly proportional to CO concentration. At CO mixing ratios of 1–2 ppm, the uptake rate was 12–120 kg/km$^2$ of land surface/day, which is equivalent to an annual total CO consumption of $7 \times 10^{14}$ to $70 \times 10^{14}$ g yr$^{-1}$. In contrast to this, Seiler and Giehl

(1976) demonstrated by experiments that various types of plants can produce CO under natural conditions. They estimated that, worldwide, plants produced $2.5 \times 10^{13}$ to $25 \times 10^{13}$ g of CO $yr^{-1}$. On the other hand, the incorporation of CO by plants is also well established by the use of $^{14}CO$ as a tracer. Therefore we must assume that plants can both remove and produce CO simultaneously, yielding a net flux from plants into the atmosphere. At present, little is known about the influence of plants on the flux of $NO_2$, $H_2O_2$, and NO.

## D.   REMOVAL AT THE SOIL SURFACE

Atmospheric trace gases can be removed at the soil surface by adsorption and microbiological processes. Adsorption is of great importance for reactive gases such as $SO_2$ and $NH_3$, which are transformed rapidly by heterogeneous reactions into sulfate and nitrate. Microbiological processes at the soil surface or in the upper layer of the soil represent a potential sink for a variety of natural and pollutant gases. *In situ* experiments have shown under natural conditions only CO to be consumed by microorganisms (Liebl, 1971; Seiler, 1974; Inman *et al.*, 1971; Liebl and Seiler, 1975). The consumption rates were found to be dependent on the type of soil and soil temperature. The deposition velocities ranged from $2 \times 10^2$ to $9 \times 10^2$ cm $sec^{-1}$ with an average of $4 \times 10^{-2}$ cm $sec^{-1}$; this rate yields a world uptake of $5 \times 10^{14}$ g $yr^{-1}$, a value that is about 50 percent of the total worldwide CO removal, including destruction by photochemical processes. In other experiments, the *equilibrium* concentration of CO over soil was found to be dependent on the type of soil and soil temperature. These results indicate that CO is not only consumed but also produced by microorganisms in the soil. The equilibrium concentrations reached values up to 0.1 ppm at soil temperatures higher than 40°C. Consequently, we cannot exclude the possibility that parts of the deserts and tropics may act as a net source of CO rather than a sink. Further investigations on the influence of soil on CO in arid climates and in the tropics are urgently needed.

## E.   GAS EXCHANGE ACROSS THE TROPOPAUSE

Tropospheric gases enter the stratosphere through transport across the tropopause. In the stratosphere, they can be destroyed by various reactions, the best understood of which are photochemical processes. The net flux of gases is dependent on the transport properties of the air and the concentrations of gases at the tropopause. For CO, the net flux is calculated to be $1.1 \times 10^{14}$ g yr$^{-1}$, which is about 10 percent of the total CO destruction. Because of the higher CO mixing ratios in the northern hemisphere (0.15 ppm) relative to the southern hemisphere (0.06 ppm), the CO flux into the stratosphere occurs mainly in the northern hemisphere ($0.9 \times 10^{14}$ g yr$^{-1}$). Unfortunately, complete profiles of $NO_2$, $NO$, $NH_3$, $HNO_3$, and $SO_2$ have not yet been measured so that the influence of the air exchange at the tropopause cannot be calculated; however, the low concentrations of these species in the troposphere and their relatively short lifetime (varying between a few seconds and a few days) make it reasonable to assume that this removal process is very small compared with other removal processes in the troposphere.

### F. GAS EXCHANGE ACROSS THE AIR–SEA INTERFACE

All gases that are soluble in water are exchanged across the air–sea interface with a net flux into the ocean. The exchange rate of an individual gas is dependent on several parameters such as the solubility coefficient, pH values, transport processes in the liquid and gaseous phases, and the gas concentration in the uppermost layer of the ocean and the marine atmosphere. Recently, models of transport through a hypotheoretical surface microlayer have been developed (Broecker and Peng, 1974; Liss and Slater, 1974) to calculate flux rates into and out of the oceans. Using CO concentrations in seawater measured by Swinnerton et al. (1974) and Seiler (1974), it is calculated that CO is released from the ocean into the atmosphere with a flux rate of $0.4 \times 10^{14}$ g yr$^{-1}$. Values of $70 \times 10^{12}$ g yr$^{-1}$ were calculated for the flux of dimethyl sulfide from the ocean to the atmosphere (Liss and Slater, 1974). In contrast to these gases, the oceans act as a strong sink for $SO_2$, the removal rate of which is calculated to be $25 \times 10^{12}$ to $75 \times 10^{12}$ g yr$^{-1}$.

## VI. RECOMMENDATIONS

Many of the panel's recommendations are either explicitly stated or implied in the discussions in the preceding sections. The rationale and methodology for measurements of the various trace species of interest are summarized in Table 8.4. In effect, this table contains the recommendations of the panel regarding the types of measurements that are required and the methods that must be developed to make them. A major point brought out in the panel's deliberations is the high degree of desirability of obtaining simultaneous measurements of these interrelated species. This concept obviously is extended to include measurements of particulate matter—size distributions, composition, and concentrations. It was noted that most, if not all, of the past programs designed to measure trace substances in the atmosphere have been limited to one, or a few, species at a time. Clearly, it will not be feasible to measure all the compounds with the desired frequency. However, in view of (i) the complexity of the various atmospheric processes, (ii) the need to simulate these processes for environmental planning and control, and (iii) the relative lack of knowledge concerning distributions and removal mechanisms of many species, the panel recommends the following:

*Existing atmospheric chemical programs should strive to incorporate the simultaneous measurement of as many of the species discussed herein as is possible within the framework of the program. These include, in the gas phase, $SO_2$, $H_2S$, $CH_3SH$, $CH_3(2)S$, $NO$, $NO_2$, $HNO_3$, $NH_3$, $O_3$, $OH$, $H_2O_2$, and $CO$; in the aerosol phase, $SO_x^{-2}$ and $NO_3^{-1}$.*

*Future atmospheric measurements programs should be specifically designed to include simultaneous measurements of as many of these species of interest to atmospheric scientists as possible.*

Such programs would involve the close cooperative efforts of several, or many, investigators in a manner that heretofore has not been common among atmospheric chemists. Yet the advantages seem obvious. Such cooperation would increase cost effectiveness in the use of expensive platforms such as aircraft, balloons, ships, and satellites. Data quality control measures could be more easily developed and implemented. Finally, integrated experiments would more clearly establish whether correlations exist among the various species. At present, many models assume specific correlations to exist even though there is essentially no evidence to support such assumptions. Thus, data from cooperative experiments would constitute an important aspect of model verification.

Complete understanding of the atmospheric chemical system discussed in this chapter will require a knowledge of concentration distributions and their temporal and spatial variability over land and over the oceans. A measurements program focused on the maritime atmosphere in relatively remote regions of the ocean would provide valuable information on many fundamental aspects of the chemical system in that such regions are relatively free of significant pollution sources.

TABLE 8.4 Measurements of Nonhydrocarbon Gases

| Type of Species | Why Measure | Method | Appropriate for Network Monitoring | Remarks |
|---|---|---|---|---|
| $SO_2$ | 1. Strong influence on global aerosol composition and distribution<br><br>2. Direct biological effects | 1. Flame photometric<br>2. Conventional resonance fluorescence<br>3. Laser-induced fluorescence<br>4. Conductivity<br>5. Uv laser backscatter<br>6. Colorimetric analysis | Yes<br><br>Yes<br><br>No<br>Yes<br>No | 1. Methods 1, 2, 3, 4, and 6 all capable of measuring $SO_2$ at pollution levels<br>2. Methods 3 and 6 capable of measuring natural levels of $SO_2$. Methods 3 and 6 are still under development<br>3. Other methods for making measurements of natural background levels of $SO_2$ should also be pursued |
| $H_2S$<br>$CH_3S$<br>$(CH_3)_2S$<br>$CH_3SH$ | 1. Via atmospheric oxidation processes these gases are significant natural sources of $SO_2$ | 1. Gas chromatography | No | Further research needed |
| NO | 1. Of major importance in determining tropospheric ozone budgets | 1. Chemiluminescence | Yes | 1. Capable of detecting natural levels at some altitudes<br>2. New research still needed |
| $NO_2$ | 1. Of major importance in determining tropospheric ozone budgets | 1. Chemiluminescence<br>2. Long-path infrared<br>3. Long-path uv absorption<br>4. Laser-induced fluorescence | Yes<br>No<br>No<br><br>No | 1. All four methods capable of measuring pollution levels<br>2. Methods 3 and 4 and possibly 1 capable of detecting natural levels<br>3. Method 4 still under development |
| $HNO_3$ | 1. Major sink for nitrogen oxides; hence, indirectly affects ozone budget | None known | — | New research needed |

206

| | | | | |
|---|---|---|---|---|
| | 2. A cause of acid rain[?]<br>3. Biological effects | | | |
| $NH_3$ | 1. Plays key role in the formation of sulfate aerosols | No method fully developed for detection of natural levels | — | New research badly needed |
| $O_3$ | 1. Plays a major role in controlling the rate of production of the free-radical OH<br>2. Biological effects<br>3. Possibly important in the heterogeneous conversion of $SO_2$ to sulfate | 1. Chemiluminesence<br>2. Uv absorption<br>3. KI | Yes<br>Yes<br>Yes | All methods capable of measuring $O_3$ at both pollution and natural levels |
| OH | 1. A major reactant responsible for the conversion of $SO_2$ to sulfuric acid vapor<br>2. A major reactant responsible for the conversion of natural sulfides to $SO_2$<br>3. A major reactant responsible for the chemical transformation of several natural and pollutant gases (i.e., CO, $CH_4$, $CH_3Cl$, $NO_2$, and $CH_xF_yCl_z$) | 1. Laser-induced fluorescence | No | Only limited data now available<br>Much more extensive data required |
| $H_2O_2$ | 1. Possibly important in the heterogeneous conversion of $SO_2$ to sulfates | None known | — | New research needed |
| CO | 1. Appears to now be a major controlling agent for OH radicals | 1. Gas chromatography<br>2. Mercury oxide methods<br>3. Long-path laser absorption | No<br>Yes<br>No | 1. Method 3 still under development<br>2. CO now appears to be an ideal tracer gas for examining atmospheric transport |

Other recommendations of a somewhat more general nature are briefly listed here.

A. Theoretical and laboratory studies are required of the following:

1. The rate constants for pertinent elementary homogeneous chemical reactions of the above-mentioned gases;

2. The formation of nuclei by heteromolecular chemical reactions;

3. The heterogeneous oxidation of $SO_2$ in solutions;

4. The heterogeneous oxidation of $SO_2$ in gas–solid systems relevant to the atmosphere.

B. Support should be provided for measurement programs and studies to aid in perfecting knowledge of the global cycles of sulfur, nitrogen, and carbon compounds. Detailed tropospheric profiles of the OH radical would be very useful.

C. Encouragement should be given to theoretical studies and measurement programs, designed to provide verification for models and to improve modeling capabilities for the transport of reacting chemical species.

## REFERENCES

Ba Cuong, N. B. Bonsang, and G. Lambert (1974). The atmospheric concentration of sulfur dioxide and sulfate aerosols over antarctic, subantarctic areas and oceans, *Tellus 26*, 241–248.

Bidwell, R. G. S., and D. E. Fraser (1972). Carbon uptake and metabolism by leaves, Can. J. Bot. *50*, 1435–1439.

Breeding, R. J., J. P. Lodge, J. B. Pate, D. C. Sheesley, H. B. Klonis, B. Fogle, J. A. Anderson, T. R. Engelert, and P. L. Haagenson (1973). Background trace gas concentration in the Central U. S., *J. Geophys. Res. 78*, 7057–7064.

Brinkmann, W. L. F., and U. de M. Samtos (1974). The emission of biogenic hydrogen sulfide from Amazonian flood plain lakes, *Tellus 26*, 261–267.

Broecker, W. S., and T. H. Peng (1974). Gas exchange rates between air and sea, *Tellus 26*, 21–35.

Chang, J. S., and D. J. Wuebbles (1976). A 2-D trospospheric OH model, Lawrence Livermore Report, in preparation.

Cox, R. A., and F. J. Sandalls (1974). The photo-oxidation of hydrogen sulphide and dimethyl sulphide in air, *Atmos. Environ. 8*, 1269.

Crutzen, P. J. (1974). Estimates of possible variations in total ozone due to natural causes and human activities, *Ambio 3*, 201–210.

Crutzen, P. J. (1976). Upper limits in atmospheric ozone reductions following increased application of fixed nitrogen to the soil, *Geophys. Res. Lett. 3*, 169–172.

Crutzen, P. J., and N. Fishman (1977). Average concentrations of OH in the troposphere and the budgets of $CH_4CO$, $H_2$, and $CHCCl_3$, *Geophys. Res. Lett. 4*, 321–324.

Davis, D. D., and G. Klauber (1975). Atmospheric gas phase oxidation mechanisms for the molecule $SO_2$, Proceedings of the Symposium on Chemical Kinetics Data for the Upper and Lower Atmosphere, *Int. J. Chem.-Kinetics*, p. 543.

Davis, D. D., W. Heaps, and T. McGee (1976). Direct measurements of natural tropospheric levels of OH via an aircraft-borne tunable dye laser, *Geophys. Res. Lett. 3*, 331–333.

Fischer, K., and W. Seiler (1975). CO production by higher plants [in German], in Proceedings of the 9. Internationale Tagung über die Luft-verunreinigung und Forstwirtschaft's, Marienbad/CSSR, October 15–18, Publ. in Proc., pp. 61–68.

Fonselius, S. H. (1972). *Hydrography of the Baltic Deep Basins III*, Fishery Board of Sweden, Series Hydrography, Rep. No. 13, 40 pp.

Friend, J. P. (1973). Global sulfur cycle, in *Chemistry of the Lower Atmosphere*, S. I. Rasool, ed., Plenum, New York, p. 188.

Fuchs, N. A. (1964). *The Mechanics of Aerosols*, Pergamon Press, Macmillan Co., New York.

Georgii, H. W. (1970). Atmospheric sulfur budget, *J. Geophys. Res. 75*, 2365–2371.

Granat, L., H. Rodhe, and R. O. Hallberg (1976). The global sulphur cycle, in *Nitrogen, Phosphorous and Sulfur—Global Cycles*, SCOPE Rep. 7, B. H. Svensson and R. Soderlund, eds., Ecol. Bull., Stockholm *22*, 89–134.

Hahn, J. (1974). The North Atlantic Ocean as a source of atmospheric $N_2O$, *Tellus 26*, 160–168.

Hahn, J., and C. E. Junge (1977). Atmospheric nitrous oxide; a critical review, *Z. Naturforsch.*, in press.

Healy, T. V., H. A. C. McKay, A. Pilbeam, and D. Scargill (1970). Ammonia and ammonium sulfate in the troposphere over the United Kingdom, *J. Geophys. Res. 75*, 2317–2321.

Hutchinson, G. L., R. J. Millington, and D. B. Peters (1972). Atmospheric ammonia: absorption by plant leaves, *Science 175*, 771–772.

Inman, R. E., R. B. Ingersoll, and E. A. Levy (1971). Soil: A natural sink for carbon monoxide, *Science 172*, 1229–1231.

Kiang, C. S., and D. Stauffer (1973). Chemical nucleation for various humidities and pollutants, *Faraday Symp. 7* (Fogs and Smokes), 26.

Junge, C. E. (1956). Recent investigations in air chemistry, *Tellus 8*, 127–139.

Junge, C. E. (1960). Sulfur in the atmosphere, *J. Geophys. Res. 65*, 227–237.

Junge, C. E., and T. Ryan (1958). Study of the $SO_2$ oxidation in solution and its role in atmospheric chemistry, *Quart. J. R. Meteorol. Soc. 84*, 46.

Levy, H., II (1971). Normal atmosphere: large radical and formaldehyde concentrations predicted, *Science 173*, 141–143.

Liebl, K.-H. (1971). The Soil as a Sink and a Source for Atmospheric CO, Master's Degree Thesis, Max-Planck-Institute for Chemistry and University Mainz.

Liebl, K.-H., and W. Seiler (1975). CO and $H_2$-destruction at the soil surface, Symposium on Microbiol. Production and Utilization of Gases (CO, $CH_2$, $H_2$), Gottingen.

Liss, P. S., and P. G. Slater (1974). Flux of gases across the air–sea interface, Nature *247*, 181–184.

Liu, S. C., R. J. Cicerone, T. M. Donahue, and W. L. Chameides (1976). Limitation of fertilizer induced ozone reduction by the long lifetime of the reservoir of fixed nitrogen, *Geophys. Res. Lett. 3*, 157–160.

Lodge, J. P., and J. B. Pate (1966). Atmospheric gases and particulates in Panama, *Science 153*, 408–410.

Lodge, J. P., P. A. Machado, J. B. Pate, D. C. Sheesly, and A. F. Wartburg (1974). Atmospheric trace chemistry in the American humid tropics, *Tellus 26*, 250–253.

Lovelock, J. E., R. J. Maggs, and R. A. Rasmussen (1972). Atmospheric dimethyl sulfide and the natural sulfur cycle, *Nature 239*, 452–453.

Machta, L. (1971). The role of the oceans and biosphere in the carbon dioxide cycle, *Proceedings of Nobel Symposium 20, Changing Chemistry of the Oceans*, Gothenburg, Sweden, Aug.

MacIntyre, F. (1970). Why the sea is salt, *Sci. Am. 223*, 105–115.

Manabe, S., and R. T. Wetherald (1975). The effects of doubling the $CO_2$ concentration on the climate of a general circulation model, *J. Atmos. Sci. 32*, 3–15.

Maroulis, P. J., and A. R. Bandy (1977). Estimate of the contribution of biologically produced dimethyl sulfide to the global sulfur cycle, *Science 196*, 647–648.

Maroulis, P. J., A. L. Torres, and A. R. Bandy (1977). Atmospheric concentration of carbonyl sulfide OCS in the southwestern and eastern U. S., *Geophys. Res. Lett. 4*, 510–512.

Matthews, W. H., W. W. Kellogg, and G. D. Robinson, eds. (1971). *Inadvertent Climate Modification, Study of Man's Impact on Climate (SMIC)*, MIT Press, Cambridge, Mass.

McConnell, J. C., M. B. McElroy, and S. C. Wofsy (1971). Natural sources of atmospheric CO, *Nature 233*, 187–188.

McElroy, M. B., J. W. Elkins, S. C. Wofsy, and Y. L. Yung (1975). Sources and sinks for atmospheric $N_2O$, *Rev. Geophys. Space Phys. 14*, 143–150.

Natusch, D. F. S., H. B. Klonis, H. D. Axelrod, R. J. Tack, and J. P. Lodge (1972). Sensitive methods for measurement of atmospheric hydrogen sulfide, *Anal. Chem. 44*, 2069–2070.

Noxon, J. F. (1975). Nitrogen dioxide in the stratosphere and troposphere measured by ground based absorption spectroscopy, *Science 189*, 549.

Ostlund, H. G., and J. Alexander (1963). Oxidation rate of sulfide in sea water, a preliminary study, *J. Geophys. Res. 68*, 3995–3997.

Rasmussen, R. A., and D. Pierotti (1977). Global and regional $N_2O$ measurements, presented at International Symposium on the Influence of the Biosphere upon the Atmosphere, Mainz, Germany, July 11–14 (submitted to *Pure Appl. Geophys.*).

Rasool, S. I., ed. (1973). *Chemistry of the Lower Atmosphere*, Plenum Press, New York.

Ripperton, L. A., L. Kornreich, and J. B. Worth (1970). $NO_2$ and nitric oxide in non-urban air, *J. Air Poll. Control Assessment 20*, 589–592.

Robinson, E., and R. C. Robbins (1967). Sources, abundance and fate of gaseous atmospheric pollutants, Stanford Res. Inst. Project PR-6755, prepared for the American Petroleum Institute, New York.

Robinson, E., and R. C. Robbins (1970). Gaseous nitrogen compound pollutants from urban and natural sources, *J. Air Pollut. Control Assoc. 5*, 303–306.

Schutz, K., C. Junge, R. Beck, and B. Albrecht (1970). Studies of atmospheric $N_2O$, *J. Geophys. Res. 75*, 2230–2246.

Seiler, W. (1974). The cycle of atmospheric CO, *Tellus 26*, 116–135.

Seiler, W. (1975). Carbon monoxide in the atmosphere and oceans, unpublished manuscript.

Seiler, W., and H. Giehl (1976). The influence of plants on atmospheric carbon monoxide, *Geophys. Res. Lett. 4*, 329–332.

Seiler, W., and U. Schmidt (1974). Dissolved nonconservative gases in seawater, in *The Sea V*, E. D. Goldberg, ed., Wiley-Interscience, New York, pp. 219–243.

Seiler, W., and P. Warneck (1972). Decrease of carbon monoxide mixing ratios at the tropopause, *J. Geophys. Res. 77*, 3204–3214.

Seiler, W., and H. Zankl (1976). Man's impact on the atmospheric CO-cycle, Proceedings of the Symposium on Environmental Biogeochemistry, Burlington, Ontario, Canada, Vol. 1, pp. 25–37.

Slatt, B. J., D. F. S. Natusch, J. M. Prospero, and D. L. Savoie (1977). Hydrogen sulfide in the atmosphere of the northern equatorial Atlantic Ocean and its relation to the global sulfur cycle, *Atmos. Environ.*, in press.

Sundquist, E. T., and J. R. Southam (1976). Chemical response of the oceans to fossil fuel $CO_2$, Abstracts, Geological Society of America, 1976 Ann. Mtg., Boulder *8*, 1129.

Svensson, B. H., and R. Soderlund, eds. (1976). *Nitrogen, Phosphorous and Sulphur— Global Cycles*, SCOPE Rep. 7, Ecol. Bull. Stockholm *22*.

Swinnerton, J. W., R. A. Lamontagne, and V. J. Linnenbom (1974). Carbon monoxide in the ocean environment, *Tellus 26*, 136–142.

Sze, N. D. (1977). Anthropogenic CO emissions: implications for the atmospheric $CO-OH-CH_4$ cycle, *Science 195*, 673–675.

Sze, N. D., and H. Rice (1976). Nitrogen cycle factors contributing to $N_2O$ production from fertilizers, *Geophys. Res. Lett. 3*, 343–346.

Tsunogai, S. (1971). Ammonia in the oceanic atmosphere and the cycle of nitrogen compounds through the atmosphere and the hydrosphere, *Geochem. J. 5*, 57–67.

Weinstock, B., and T. Y. Chang (1974). The global balance of carbon monoxide, *Tellus 26*, 108–115

World Meteorological Organization (1975). *The Physical Basis of Climate and Climate Modelling*, Rep. Internat. Study Conf., Stockholm, July–August 1974, GARP Pub. Ser. No. 16, WMO-ICSU.

# 9 Radionuclides

## I. INTRODUCTION

The troposphere is an important medium for the transport of natural and man-made radionuclides to the oceans. At this time, the principal sources of the man-made radionuclides present in the atmosphere are the detonations of nuclear devices and the releases from various operations in the nuclear-power fuel cycle (mining, fuel fabrication, reactor operation, fuel reprocessing, and waste management). These include a number of well-studied isotopes such as $^{90}$Sr, $^{137}$Cs, $^{14}$C, and $^{3}$H, as well as the less thoroughly investigated transuranic species such as plutonium, americium, and curium and noble gases such as $^{85}$Kr. The natural radionuclides found in the atmosphere are primarily members of the uranium decay series, in particular, $^{222}$Rn and its radioactive decay (daughter) products and, to a lesser degree, isotopes of the thorium decay series.

Various artificial and natural radionuclides have been extensively used as tracers of atmospheric transport processes. However, *most of the nuclear-weapons-derived material that finds its way to the sea was initially released in the stratosphere and subsequently entered the troposphere from above; these nuclides are beyond the purview of this panel.* In constrast, nuclear-fuel-cycle materials and naturally occur-

---

Members of the Working Group on Radionuclides were H. L. Volchok, *chairman*; K. K. Turekian, E. A. Martell, and S. Tsunogai.

212

ring radioisotopes are injected into the atmosphere from the earth's surface; therefore, their atmospheric route to the oceans must be almost exclusively via the troposphere. *In the subject area of nuclear power and fuel-cycle sources, we have excluded from our consideration the analysis of the consequences of catastrophic incidents (such as might occur in a nuclear reactor excursion); this subject has been thoroughly studied and reported on in the literature.*

## II. ARTIFICIAL RADIONUCLIDES

A great deal of research has been carried out on the sources and distributions of artificial radionuclides in the earth's environment, particularly with respect to the fission products and transuranic elements associated with nuclear weapons tests (e.g., United Nations, 1972). In general, radionuclides deposited as aerosols on continental surfaces appear to have very long residence times in soil. These aerosols (as well as $^3H$ and $^{14}C$, which occur primarily in gaseous form in the atmosphere) also have very long residence times in the oceans.

In the nuclear fuel cycle, there are numerous opportunities for radioactive releases to the atmosphere (Volchok *et al.*, 1974). The release can occur as a consequence of accidents or as routine operational "leakages"; the released material can be in the gaseous state or the solid (particulate) state. The principal radionuclides in the fuel enrichment and fabrication processes are the isotopes of uranium, thorium, and plutonium; releases of these materials would occur as particulates. As far as reactor operations are concerned, facilities are designed in such a manner that the release of particulates is precluded in routine operation; however, gaseous species such as $^3H$ (tritium), $^{131}I$, and $^{85}Kr$ can be routinely emitted in substantial quantities (Preston *et al.*, 1972).

Tritium (halflife = 12.26 years) from the reactor fuel cycle is released as tritium gas ($T_2$) or tritiated hydrogen (HT and DT); in contrast, the tritium produced in fusion weapons tests occurs as HTO. The other major source of the tritium gas release is the venting from underground testing of large fusion devices (Mason and Ostlund, 1974). At present, approximately 95 percent of the atmospheric inventory of tritium is weapons-produced. There is no evidence that the present (or the projected future) concentration of tritium in the atmosphere will have any significant effects on physical or biological systems. On the other hand, tritium is a useful tracer for the study of atmospheric and oceanic transport and mixing processes having characteristic times of months to years; thus, tritium could provide insights into the behavior of some types of pollutants.

Iodine-131, because of its short halflife (8.08 days), should be of no signficance as far as transport to the oceans is concerned.

Kyrpton-85 (halflife = 10.76 years) is produced both in the nuclear-power fuel cycle and in fusion bomb testing. However, in contrast to tritium, more than 95 percent of the $^{85}$Kr that is present in the environment on a global scale originates from the power cycle, and only 5 percent is from weapons testing (Schroder and Roether, 1974). The projected increase in concentration of $^{85}$Kr will lead to a significant increase in air ionization, 43 percent above the present level by the early part of the next century (Boeck, 1976); because atmospheric conductivity is proportional to air ionization, the conductivity will increase by the same amount. There are several pathways by which air ionization due to $^{85}$Kr could have an environmental impact (Boeck, 1976): by influencing storm electrification; by affecting the coalescence of cloud droplets to form raindrops, especially in cloudbursts; and by affecting the formation of sulfate aerosol particles in gas-to-particle conversion mechanisms. However, at present, there is no firm evidence linking $^{85}$Kr to atmospheric phenomena.

No direct measurements of the tropospheric transport of artificial radionuclides from land to the sea have been reported. However, some site-specific plutonium isotope anomalies have been noted in a few local coastal marine sediment environments; these may reflect an atmospheric transport mechanism (Koide et al., 1975). Direct measurement of the flux via this route is difficult because of the obscuring effects of the fallout of these same nuclides from the dominant source—atmospheric nuclear detonations, most recently those of France and the People's Republic of China. These explosions deposit most of the radioactive debris in the stratosphere; this material is subsequently transferred to the troposphere by a number of mechanisms at a rate that yields a stratospheric half-residence time (between stratosphere and troposphere) of about 10 months.

The concentration in tropospheric air of the artificial radionuclides derived directly from the stratosphere is undoubtedly much greater than their concentrations due to resuspension of soil containing material deposited earlier. This can be clearly shown by considering reasonable upper limits of resuspension concentrations by use of the "resuspension factor." This is a parameter that relates air concentrations of a substance to its areal deposit within a specific geographical region. It is defined as the concentration per unit volume of air divided by the quantity contained per unit area of soil; in the case in point, for radioactive substances,

$$\text{Resuspension Factor (m}^{-1}) = \frac{\text{mCi/m}^3 \text{ in air}}{\text{mCi/m}^2 \text{ in soil}},$$

where mCi = $10^{-3}$ curie.

Consider the situation in the vicinity of Rocky Flats, Colorado, where an accident in a processing plant contaminated a fairly large soil area with plutonium. The Rocky Flats area is basically desert, with mean annual precipitation of about 40 cm and mean wind velocity of some 13 km per hour. The ground cover is typical desert grass and cactus, and the soil in the vicinity of the contamination is quite sandy and unconsolidated. The climatology, vegetation, and soil characteristics are such as to lead us to expect resuspension at this site. The resuspension factor of Rocky Flats is about $10^{-9}$ m$^{-1}$ (Volchok, 1971; Krey *et al.*, 1974). In plowed areas or other soil regions affected by human activity, the factor might be considerably higher.

If, for illustrative purposes, we apply this factor to the entire 40° to 50° latitude band of the northern hemisphere, where the mean $^{239+240}$Pu (hereafter referred to as $^{239}$Pu) deposit is about 2 mCi/km$^2$ (Hardy *et al.*, 1972), a resuspension air concentration of 2 aCi/m$^3$ is predicted (aCi = $10^{-18}$ Ci). The actual measured mean annual air concentration in this latitude band in 1974 was about 50 aCi/m$^3$ (Volchok *et al.*, 1975). On this basis, it would appear that resuspended soil containing $^{239}$Pu could, at most, account for just a few percent of the total surface air concentration and that the resuspended input will be undetectable until all atmospheric nuclear testing is stopped and the stratospheric reservoir has been substantially depleted.

An interesting sidelight to the Rocky Flats study is that an effort was made to delineate the extent of the affected area. The most sensitive analytical system (mass spectrometric analyses of Pu isotopes) was employed to measure the soil concentrations of the spill material. The maximum downwind distance at which redistributed Pu from the spill area could be identified was 30 km (Krey, 1976).

We have also compared surface-air concentration data from sites of comparable latitude but of contrasting climate. Examples are given in Table 9.1. These data suggest that air concentrations are not significantly higher in areas of relatively low rainfall, where the soil is drier and supposedly more mobile and, hence, where one might expect resuspension to be at a maximum.

Little research has been done on describing quantitatively the "resuspendable layer" of the soil surface. Thus, we do not know at what rate, and to what depth, deposited radionuclides are transported down the soil column before they become unavailable for resuspension (Hardy, 1974).

TABLE 9.1    Plutonium-239 Air Concentrations[a]

| Year and Location | Latitude | Precipitation (cm) | $^{239}$Pu (aCi/m$^3$)[c] |
|---|---|---|---|
| 1970 | | | |
| Richland, Wash. | 45° N | 15 | 53[b] |
| New York, N.Y. | 41° N | 90 | 68 |
| 1974 | | | |
| Richmond, Calif. | 38° N | 47 | 26 |
| New York, N.Y. | 41° N | 121 | 39 |
| Helena, Mont. | 47° N | 27 | 42 |

[a] Data from Volchok et al. (1975) except [b]Thomas and Perkins (1974).
[c] aCi/m$^3$ ≡ 10$^{-18}$ curie.

Based on the foregoing, it is apparent that the tropospheric transport of man-made radionuclides into the sea from material deposited on land surfaces is, on a global scale, a relatively small fraction of that being delivered from recent nuclear tests and that until these tests substantially cease it will be difficult to quantify this mode of oceanic supply.

## III.   NATURAL RADIONUCLIDES

Of the naturally occurring radionuclides, $^{222}$Rn and $^{210}$Pb are potentially the most useful for the study of transport between land and sea. Radon-222, with a 3.8-day halflife, is emanated from soils (and, also, from mining operations) and ultimately decays to $^{210}$Pb (halflife = 22 years; see Figure 9.1 for the $^{226}$Rn decay series). Radon-222 moves as a gas by advection and diffusion in the air column. When $^{222}$Rn decays, its decay products ($^{210}$Pb and its precursors) rapidly become attached to aerosols, which are then subject to removal by precipitation and as dry fallout. This behavior of $^{222}$Rn and its $^{210}$Pb daughter is analogous to the behavior of other trace gases and their reaction products injected into the atmosphere (for example, $SO_2$ and its oxidation product $SO_4^{2-}$). However, radon is itself an inert gas. It is evident that the $^{222}$Rn/$^{210}$Pb system can serve as a tracer as well as a clock for the study of such reactions and atmospheric concentrations. Also, the study of the behavior of $^{210}$Pb will provide insights into the expected behavior of other particulate trace substances of similar size in the atmosphere and may offer a basis for inferring the behavior of particles of other sizes. (The geochemistry of radon and its daughter products is discussed in Turekian et al., 1976.)

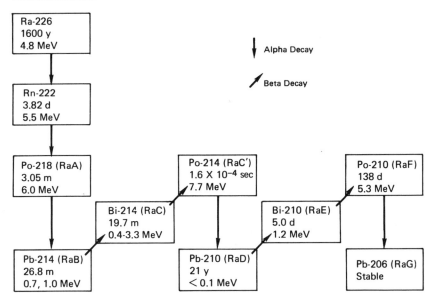

FIGURE 9.1 Principal decay scheme of the radium-226 series as of August 1975.

## A. ATMOSPHERIC RADON MEASUREMENTS

It has been a common practice in the past to measure the concentration of the short-lived daughters of radon rather than concentration of radon itself; this was done for reasons of convenience and economics because of the relative ease with which the particulate-phase daughters could be collected and counted on a filter. However, under conditions of high humidity and high aerosol concentrations, there is a danger that the concentration of the short-lived daughters may be considerably less than the expected radioactive equilibrium concentration with the ambient radon. In such cases, it is evident that the measurement of daughters cannot be assumed to be a reliable measure of the radon concentration. Because non-clear-air conditions are more common than clear-air conditions in surface air, it is more reliable to measure radon directly whenever possible.

## B. RADON FLUX FROM SOILS

The radon flux from soils is much greater (approximately $10^2$ times greater) than that from the ocean surface (Wilkening and Clements, 1975); however, an accurate knowledge of the geographical and time dependence of the radon flux from the continents and oceans is needed both for air-parcel tracing and for continuity arguments involving the atmospheric $^{210}$Pb flux. The flux of radon emanation from soil has been measured or estimated by four methods to date: (1) direct measurement of radon in an enclosed container lying on the soil surface or ocean surface (Wilkening and Clements, 1975); (2) measurement of short-lived daughter activity above soils using nuclear detectors and dosimeters; (3) modeling of the radon flux from $^{210}$Pb and $^{226}$Ra measurements in old soil profiles (Turekian et al. 1976); and (4) calculation of the average worldwide radon flux from models of atmospheric $^{210}$Pb flux distributions on the basis of continuity arguments (Turekian et al., 1976).

Geographic and seasonal radon flux data of adequate quality and sufficient quantity are lacking, and major disparities exist among the limited data available. We do know that the number of condensation nuclei above vast snow-covered areas is small and, thus, that the radon flux should also be small. We also know that the quantity of radon and of short-lived radon daughter products over soils are affected by soil properties and meteorological conditions; however, on a large scale, there appears to be a reasonable degree of homogenization of radon fluxes. Clearly, more measurements of the radon flux over soils are necessary to resolve these questions and to provide adequate source terms for tracer studies of air parcels.

## C.  LEAD-210 IN THE ENVIRONMENT

The critical problems regarding atmospheric $^{210}$Pb are the mean residence time, generally expressed in days, and the geographic distribution of the atmospheric flux of $^{210}$Pb, generally expressed as pCi per m$^2$ per year (pCi = $10^{-12}$ Ci). Because the dominant source of $^{210}$Pb in the atmosphere is continentally derived radon, the knowledge of these two parameters could provide some insight into the atmospheric survival times of other continentally derived substances as the air parcels move over the oceans; also, these substances could serve as a tracer for distinguishing between air parcels that have had a relatively long residence time and those that have recently emerged from the continents.

## D.  METHODS OF MEASURING LEAD-210 FLUX

There are four methods of measuring the atmospheric $^{210}$Pb flux: (1) the actual collection and analysis of fallout, both wet and dry, over a long period of time; (2) the measurement of the $^{210}$Pb concentration in the air column and the subsequent application of the appropriate deposition velocity (Moore *et al.*, 1973); (3) the measurement of $^{210}$Pb and $^{226}$Ra in repositories such as soil profiles and other appropriate integrators (Turekian *et al.*, 1976); and (4) from the distribution of $^{210}$Pb and $^{210}$Bi concentrations and activity ratios as a function of altitude (Moore *et al.*, 1973).

There is a paucity of flux data. The fallout method of obtaining flux, while most direct, has produced few acceptable data because of the requirements for continuous, quantitative sampling and analysis. In the second method, atmospheric concentration data are obtained on a short time scale using air samplers on land or on airplanes, but the deposition velocity that must be applied to calculate flux has not, to date, been obtained for $^{210}$Pb in many cases. The repository technique has been used in old soils, glaciers, and lakes or in deposits on man-made ponds. This is a reliable method because of the quantitative retention of $^{210}$Pb in these environments, but a minor correction for near-surface radon loss in the repository must be determined or modeled to get the most accurate values in various areas.

E.   MEAN RESIDENCE TIME OF ATMOSPHERIC LEAD-210

There has been considerable discussion about the residence time of $^{210}$Pb in the troposphere as a function of season and altitude. The most commonly applicable mean residence time for tropospheric $^{210}$Pb appears to be less than one week (Martell and Moore, 1974), although there are studies indicating substantially longer values (e.g., Nozaki and Tsunogai, 1972; Tsunogai and Fukuda, 1974). More research must be done to establish the constancy of this value. In particular, it appears that in the lower troposphere over the oceans the mean residence time of aerosols may be shorter than over continents. Further work should be done in this area, including the direct measurement of deposition velocities or mean residence times by coupled atmospheric concentration profiles and surface-level measured fluxes.

There is evidence that the western parts of continents may receive higher $^{210}$Pb fluxes in some latitudes than predicted by a simple zonal model of Rn decay and $^{210}$Pb precipitation over oceans. Three-dimensional modeling (Mahlman, Princeton University, private communication) of the $^{210}$Pb distribution in the atmosphere and the flux may be useful in understanding these flux observations.

## IV. RECOMMENDATIONS

1. The system of $^{222}$Rn and its daughter products has the potential of being extremely useful as a tracer for the validating of atmospheric transport models that incorporate a particulate phase in the removal processes. To this end, the vertical and areal distribution of Rn and of $^{210}$Pb must be better characterized over the continents and the oceans, especially the latter.

2. There is some evidence that under certain environmental conditions radon is not in radioactive equilibrium with its short-lived daughter products in the aerosol phase; hence, the concentration of radon cannot always be accurately inferred from the measurement of the daughter-product activity collected on filters. Nonetheless, the filter technique is appealing because of its simplicity. Therefore, we recommend that studies be made of the relationship between radon and its daughter products under a wide range of meteorological and climatic conditions so as to assess the operational limitations of the filter technique.

3. The deposition of $^{210}$Pb (and other materials) will be strongly dependent on precipitation distribution and intensity, parameters that are poorly characterized over the oceans. Our knowledge in this area must be greatly improved.

4. The mechanisms for the mobilization of soil radioactivity (i.e., resuspension) and the subsequent transport through the atmosphere should be more thoroughly investigated. It seems advantageous to consider such studies in the context of the general problem of the mobilization of soil aerosols.

## REFERENCES

Boeck, W. L. (1976). Meteorological consequences of atmospheric krypton-85, *Science* *193*, 195–198.

Hardy, E. P. (1974). Depth Distribution of Global Fallout Sr$^{90}$, Cs$^{137}$ and Pu$^{239,240}$ in Sandy Loam Soil, USAEC Rep. HASL-286:1–2.

Hardy, E. P., P. W. Krey, and H. L. Volchok (1972). Global Inventory and Distribution of Pu-238 from SNAP-9A, USAEC Rep. HASL-250.

Koide, M., J. J. Griffin, and E. D. Goldberg (1975). Records of plutonium fallout in marine and terrestrial samples, *J. Geophys. Res. 80*, 4153–4162.

Krey, P. W. (1976). Remote plutonium contamination and total inventories from Rocky Flats, *Health Phys. 30*, 209–214.

Krey, P. W., R. Knuth, T. Tamura, and L. Toonkel (1974). Interrelations of surface air concentrations and soil characteristics at Rocky Flats, presented at the Symposium on Atmospheric Surface Exchange of Particulate and Gaseous Pollutants at Richland, Washington, Sept. 4–6.

Martell, E. A., and H. E. Moore (1974). Tropospheric aerosol residence times: a critical review, *J. Rech. Atmos. 8*, 903–910.

Mason, A. S., and H. G. Ostlund (1974). Atmospheric HT and HTO: major HT injections into the atmosphere, 1973, *Geophys. Res. Lett. 1*, 247–248.

Moore, H. E., S. E. Poet, and E. A. Martell (1973). $^{222}$Rn, $^{210}$Pb, $^{210}$Bi and $^{210}$Po profiles and aerosol residence times vs. altitude, *J. Geophys. Res. 78*, 7065–7075.

Nozaki, Y., and S. Tsunogai (1973). Lead-210 in the North Pacific and transport of terrestrial materials through the atmosphere, *Earth Planet. Sci. Lett. 20*, 88.

Preston, A., R. Fukai, H. L. Volchok, N. Yamagata, and J. W. R. Dutton (1972). Radioactivity, in *A Guide of Marine Pollution Seminar—Food and Agriculture Organization of the U.N.*, Gordon & Breach Science Publications, New York.

Schroder, K. J. P., and W. Roether (1974). The releases of krypton-85 and tritium to the environment and tritium to krypton-85 ratios as source indicators, in *Isotope Ratios as Pollutant Source and Behavior Indicators*, IAEA Symposium (IAEA-SM-191/30), Vienna, p. 231.

Thomas, C. W., and R. W. Perkins (1974). Transuranium Elements in the Atmosphere, Battelle Pacific Northwest Laboratories Rep. BNWL-1881, VC 48 (presented at the ANS meeting at Washington, D.C.).

Tsunogai, S., and K. Fukuda (1974). Pb-210, Bi-210 and Po-210 in meteoric precipitation and the residence time of tropospheric aerosol, *Geochem. J. 8*, 141.

Turekian, K. K., Y. Nozaki, and L. K. Benninger (1976). Geochemistry of atmospheric radon and radon products. In *Annual Review of Earth and Planetary Sciences*, F. A. Donath, ed., Annual Reviews, Inc., Palo Alto, Calif., Vol. 5, pp. 227–255.

United Nations (1972). *Ionizing Radiation: Level and Effects, Volume I, Levels*, A Report of the United Nations Scientific Committee on the Effects of Atomic Radiation, UN Publication, Sales No. E.72.IX.17.

Volchok, H. L. (1971). Resuspension of plutonium-239 in the vicinity of Rocky Flats, Los Alamos Symposium, LA-4756, p. 99.

Volchok, H. L. (1974). Radioactivity in the atmosphere, in *Chemist-Meteorologist Workshop*, Rep. WASH-1217-74 (Ft. Lauderdale, Fla., Jan. 14–18).

Volchok, H. L., L. Toonkel, and M. Schonberg (1975). Radionuclides and Pb in Surface Air, USERDA Rep. HASL-297 (Appendix), B-1–B-140.

Wilkening, M. H., and W. E. Clements (1975). Radon-222 from the ocean surface, *J. Geophys. Res. 80*, 3828–3830.

# 10 Techniques

## I. INTRODUCTION

In order to evaluate the transfer of chemical substances between the atmosphere and the ocean, it is necessary either to (i) measure the flux directly or (ii) measure concentrations in the atmosphere and then calculate the net flux to the ocean. Both of these approaches require the quantitative collection of representative samples of gases, airborne particles, and precipitation. The flux of material across the air–sea interface can be in either direction, but this chapter will focus on evaluating the transfer of airborne material to the ocean.

It is vital to answer several questions before a collection and measurement program can be undertaken.

1. What are the precise objectives of the study?
2. What is the structure of the concentration–deposition field?
3. What is the required spatial extent and density of sampling sites?
4. What sampling frequency and duration are required?

Answers to these questions must be provided by the experimental design of individual studies. Obviously, the requirements of a research program could differ significantly from those of a monitoring program

---

Members of the Working Group on Techniques were D. M. Whelpdale, *chairman*; and G. E. Likens.

in this regard. Some aspects of these problems have been discussed elsewhere in this work, especially in Chapters 3 and 4. In this chapter, the discussion is restricted to collection techniques for gases, particles, and hydrometeors in the atmosphere. It is our intention to give a broad overview of techniques currently in use. More detailed coverage can be obtained from the cited references.

## II. COLLECTION TECHNIQUES

The main objective of the sample-collection techniques discussed herein is to obtain information on the chemical composition and concentration of material in the atmosphere. A large variety of collection procedures, sampling devices, and analytical techniques have been utilized (see also Duce *et al.*, 1976; Windom and Duce, 1976).

### A. GASES

Many techniques with varying degrees of sophistication are available for the collection and *in situ* measurement of the concentration of several inorganic gases. Very few, however, can be used for unattended, long-term sampling.

Collection systems for several of the high-molecular-weight gaseous hydrocarbons are presently under development. Collection systems for chlorinated hydrocarbons such as PCB and DDT have been developed by Bidleman and Olney (1974) and by Harvey and Steinhauer (1974). Techniques for the determination of the chlorofluorocarbons have been described by Grimsrud and Rasmussen (1975), Su and Goldberg (1973), and Lovelock *et al.* (1973). The distribution of several atmospheric hydrocarbons, including PCB and DDT, between the gas and particulate phases is largely unknown; therefore, it is often necessary to sample both phases separately or to use a system that collects a total air sample.

### B. PARTICLES

Samples of airborne particles are collected by filtration, impaction, or a combination of the two. Filter systems usually consist of an intake nozzle, filter, flow meter, and regulator and a pump. The instrument used routinely by control and surveillance agencies is the high-volume sampler with a flow rate that is typically 1–2 $m^3$/min (Jutze and Foster, 1967). Modifications of this system and many other types are used for specialized research applications such as "clean site" monitoring at Mauna Loa (Simpson, 1972), collection of sea-salt particles (Fasching *et al.*, 1974), and simultaneous gaseous $SO_2$ and particulate $SO_4^{2-}$ sampling by aircraft (Johnson and Atkins, 1974).

Filter samples are collected for three purposes: to determine the total mass loading of particles, to conduct chemical analyses, and to study particle morphology. Sampling times of one to several days at high flow rates (typically 2 $m^3$/min) are required to collect an adequately large sample. For chemical analysis, the composition of the filter is critical (Dams *et al.*, 1972): it must contain relatively small amounts of the species being collected and it must be compatible with proposed analytical procedures. Cellulose filters (e.g., Whatman 41) are commonly used for sulfur analyses; cellulose (e.g., Whatman 41), polystyrene (e.g., Delbag Microsorban) or Nuclepore for heavy metals; and glass fiber (e.g., Gelman Type A) for organic analysis. For the collection of trace-level inorganic species, filters should be mounted in plastic filter holders (polyethylene), and the holders should be placed in plastic shelters to protect them from rain. After collection, filters should be stored in sealed polyethylene containers. For organic species, similar procedures should be followed, using an aluminum or stainless-steel collection apparatus.

Other factors determining the choice of filters include the flow rate obtainable through the filter, the efficiency of the filter as a function of particle size, loading characteristics (rate at which filter clogs), and cost. More detailed information on filter characteristics is available in Stafford and Ettinger (1972), Spurney *et al.* (1969), Butcher and Charlson (1972), and Liu and Lee (1976).

Problems encountered in filter sampling include the collection of a representative sample of the aerosol, the accurate measurement of sample volume, and the chemical stability of the sample. The entire sampling system has a characteristic upper cutoff size (determined by the geometry of the intake and the wind velocity) and lower cutoff size (related to the collection efficiency of the filter), which can result in a biased sample of the aerosol. For example, the standard high-volume system used in the United States has especially poor intake characteristics that result in sharply reduced efficiency for particles above 5-$\mu$m diameter (Wedding *et al.*, 1977); also, because of the intake configuration, the cutoff size varies with wind direction. Chemical changes in the sample can be caused by reactions among the sampled materials, by reactions with atmospheric trace gases and with water vapor (Johnson and Atkins, 1974; Forrest and Newman, 1973), and by loss of water or volatile organic species.

Large Terylene or nylon meshes mounted normal to the wind have also been used to collect particles. They may be located on land-based towers or onboard ships or flown as kites from ships (Delany *et al.,* 1967). Although semiquantitative because of the uncertain and variable collection and retention efficiencies of the mesh, the technique is attractive because of the possibility of sampling large volumes of air ($10^6$ m³/day with several square meters of mesh) and because of its simplicity, durability, and low cost. Evidence suggests that the mass collection efficiency is relatively constant for a specific aerosol type sampled in a specific meteorological setting (Prospero and Nees, 1977).

Air-sampling instruments that make use of inertial separation of particles from an airstream or inertial impaction onto a surface include impactors, impingers, and many others (ACGIH, 1972). Impactors, the most frequently used, are available in a variety of configurations: single-stage jet impactors and cascade impactors such as the Casella, Lundgren, and multijet Andersen. Recent developments in the theory of rectangular (slot) and round (jet) impactors (Marple and Liu, 1974, 1975) make it possible to design impactors with optimal operational characteristics. The relatively new high-volume cascade impactors having flow rates over 1 m³/min are particularly useful because they can sample large volumes of air. Frequently, a filter is added as the final stage to collect particles smaller than 0.3–0.5 $\mu$m that have passed through the impactor. Serious errors in the size distribution can be introduced by particle bounce-off, re-entrainment, and particle shattering at various stages; these effects can be reduced or eliminated by coating the collection surface with nondrying adhesive films such as greases and oils (Dzubay *et al.,* 1976).

A variation of the principle of operation of the impactor is incorporated in the cascade centripeter (Hounam and Sherwood, 1965) and the virtual impactor (Stevens and Dzubay, 1975); air entering the instrument through an inlet jet is directed against an "impaction" stage that consists of a slowly pumped void; "impacted" particles are subsequently collected on a filter. This adaptation reduces bounce-off and re-entrainment problems and minimizes reactions on the filter; it thus allows the collection of larger samples required for some chemical analyses.

Impactors separate particles according to aerodynamic size and thus can provide information on the particle number and mass distribution (Natusch and Wallace, 1976) and the chemical composition as a function of size. Disadvantages of the technique are that the particle cutoff size for each stage is not sharp and that long sample times are required to collect sufficient material for chemical analyses, which may lead to problems of chemical instability as with filters.

Particle collection systems should be designed to sample isokinetically so as to minimize size discrimination effects at the sampler intake (Davies, 1968). This may be facilitated by using appropriate intake nozzle geometry and by ensuring that the intake always faces into the wind. However, in practice, isokinetic conditions are difficult to realize (Belyaev and Levin, 1974; May *et al.*, 1976).

Table 10.1 contains a relatively complete list of features of most instrumentation used in the measurement of aerosol parameters.

### C.  PRECIPITATION

Two measurements are required to determine inputs of chemicals to oceans via precipitation: (i) the amount of precipitation and (ii) the concentration of material in the precipitation. Techniques to determine the amount of precipitation over land surfaces have, in general, been standardized; this has not been done for marine areas (e.g., Elliott and Reed, 1973). Techniques for the collection of samples for chemical analysis have not been standardized for either land or sea use.

Rain and snow are currently collected and analyzed by a number of different techniques in continental areas (e.g., Junge, 1963; Lodge *et al.*, 1968; Eriksson, 1959, 1960; Petrenchuk and Selezneva, 1970; Selezneva, 1974a, 1974b; Summers and Whelpdale, 1975). Unfortunately, little information is available relative to procedures (or results) for marine locations. Collectors normally have been located within 1 to 3 m of the earth's surface and consist of a collection surface attached to a storage reservoir. Components of collectors used for chemical analysis are usually constructed of polyethylene, glass, or stainless steel. Collectors on ships have been operated from the bow or superstructure when heading into the wind; however, contamination from sea spray and splash often is a serious problem.

Precipitation samples can be obtained with a variety of types of collectors. "Bulk" precipitation collectors are continuously open to the atmosphere; "wet" collection devices are functional (open) only during periods of rain or snow; "dry" collection devices are covered during periods of precipitation and are uncovered at all other times. Automatic (wet–dry) precipitation collectors incorporate a precipitation sensor that activates a mechanism that moves a cover from one receptacle to the other at the time of initiation or cessation of precipitation. Electrical power is required for automatic precipitation collectors; this might pose a serious problem in some remote areas. Bulk precipitation collectors (e.g., Likens *et al.*, 1967) have been widely used because of their simplicity, reliability, and lack of electrical power requirements. However, these collectors suffer from the disadvantages of providing no separation of the wet and dry deposition components and of being prone to local contamination because they are continuously open.

Sample volumes depend on the amount of precipitation and the surface area of the collector orifice and on the subsequent evaporative loss from the collector. Sample volumes varying from 100 ml to several liters are required for the chemical determinations that are typically carried out.

The major problems in attempting to characterize the chemical composition of continental precipitation have been related to the collection and storage of samples (Galloway and Likens, 1975) and to analytical difficulties (Paterson, 1975). The problems of collection include the following:

1. The effectiveness of the design of the collector in obtaining a representative sample of precipitation and its chemical content;

2. Geochemical or biological changes that occur within the collection container as a consequence of the presence of foreign particles either in the samples or on the filtering material through which the precipitation passes;

3. The effect of storage, including *in situ* collection time, on chemical composition;

4. Leaching, or absorption, of substances from, or to, collector surfaces;

5. The effects of sampling interval both during and between storm periods.

A number of collectors of varied design from different countries have been tested and intercalibrated in separate studies by U.S. and Canadian scientists. The results of these investigations (Galloway and Likens, 1975; Smith, 1975) were similar and may be summarized as follows.

1. Precipitation samples must exclude dry deposition if accurate information on the chemical composition of rain and snow is required.

2. Chemical input includes wet and dry deposition, and both must be assessed in biogeochemical studies.

3. A wet–dry collector of the ERDA/HASL (U.S.A.) or AAPS (Finnish) design proved to be the most reliable and useful.

4. Substantial contamination may result when glass and plastic collectors are used to sample precipitation for cationic and organic components, respectively. For organic material, collectors should be constructed from glass, stainless steel, or aluminum.

| Quantity | Why Measure? | Is a Method Available? | Appropriate for Network Operation? | Remarks |
|---|---|---|---|---|
| *Chemical Measurements* | | | | |
| Mass concentration of various constituents | Budget studies; need relation of aerosol mass to mass of gases for many different substances | Suspended particles: yes, if located and operated carefully | Yes | 1. Study of sampling location effects and intake losses needed<br>2. May be less relevant than other variables on regional or global scale |
| Concentration of aerosol-forming gases | Production step in aerosol cycle | $SO_2$, organics, $NO_x$: yes, Sulfides, $NH_3$: no | Research basis only No | Much method research imperative |
| Chemical composition as a function of size and mass as a function of size | Relevant to<br>1. Cloud and precipitation processes<br>2. Residence time<br>3. Heterogeneous reactions<br>4. Budget of trace elements and elemental cycles<br>5. Biological effects | Yes, but only for particles $> \sim 0.3\ \mu m$ (impactors)<br><br>Flame photometry for Na, other metals; perhaps S distribution<br>Scanning electron microscope with energy dispersive x-ray analyzer | No, but desirable on an experimental basis<br><br>Research basis only<br><br><br>Research only | 1. Need information on $r < 0.1\ \mu m$<br>2. What is effect of particle shape in impactors<br>3. Trace element role in heterogeneous reaction<br>4. Need indication of secular change |
| Particle solubility | Mode of action as cloud nuclei | Yes, but as a bulk property only | No | Research needed |
| Particle area as a function of size | Large role in heterogeneous reactions | Research basis only, may be questionable | No | Important research topic |

228

| Parameter | Role | Measurement status | | Research needs |
|---|---|---|---|---|
| Deposition (wet) | 1. Ecological effects and long-term effects 2. Budgets; biogeochemical cycles 3. Indication of aerosol composition (crude) | Wet removal. Measurements are reasonably accurate for long-term averages, but sampling time and logistics may need reconsideration | Yes | 1. Need research with regard to short-term variation and relation to synoptics 2. Organic analysis needed 3. Heavy metals |
| Deposition (dry) | Same comments as for wet deposition | Research stage | No, except deposition plate; no realistic way of simulating real surfaces | Location of sampler is important; altitude dependence needed |
| Particle density and shape | Role in impaction processes and in optics | Yes, partial (e.g., impactor and electron microscopes) | No | Research topic |
| $N(r)$ for $0.001 \leq r \leq 0.1\ \mu m$ | Gas-to-particle conversion products in this range | Ion mobility analyzer; Diffusion battery + condensation nucleus center | No No | Research only Needs improvements for background analyses |
| $N(r)$ for $0.1\ \mu m \leq r \leq 1\ \mu m$ | Effects on radiation budget, optical effects, cloud formation | Several; expensive or time consuming in evaluation; optical counters, optical and electron microscope, for example | No | Need better understanding of particle shape effects in optical counters |
| $N(r)$ for $1\ \mu m \leq r \leq 10\ \mu m$ | Cloud formation Mass deposition Mass budget | Yes, impactors | No | Same as above |

[a] Adapted from B. Bolin, G. Witt, and R. Charlson (1973). *Stockholm Tropospheric Aerosol Seminar* (STAS) Rep. AP-14, Inst. Meteorology, U. of Stockholm, and the International Meteorological Inst. in Stockholm, Sweden.

5. Collectors should be cleaned between periods of use. If acids are used, electrical conductivity of the final rinse solution should be monitored for contamination.

6. Concentrations for most inorganic ions in precipitation samples of low pH (3.5 to 4.5) did not change significantly when stored at 4°C for a period of 8 months. Chloride and phosphate results were variable.

7. Precipitation samples should be collected on an event basis whenever possible, or for intervals of not longer than a week, especially if the pH is greater than 5, because of the possibility of composition change as a result of biological activity.

8. Samples with visible particulate contaminants should be discarded.

9. Network stations should be readily accessible yet remote from obvious local sources of pollution (e.g., agriculture, smelters, airplanes).

10. Dry deposition is not quantitatively measured by current designs of bulk or wet–dry precipitation collectors.

The Soviet Union has an active program of precipitation collection and analysis onboard ships (Miller, 1975a; Selezneva, 1974a, 1974b). In general, however, there are relatively few data published on precipitation chemistry from shipboard collections. Gambell and Fisher (1966) report on the chemical composition of precipitation from two shipboard collections off the coast of North Carolina. Specific problems associated with sampling at sea include the following:

1. Influence of airborne continental material on marine precipitation chemistry;

2. Heterogeneity of concentration–deposition field;

3. Height of collection vessel above the sea surface, the deck of the ship, and the buoy or platform surface;

4. Contamination by, for example, birds, sea spray, or raindrop splash.

## III. GENERAL CONSIDERATIONS

### A. PHYSICAL AND CHEMICAL PROPERTIES

Frequently, chemical species exist in more than one phase in the atmosphere. In some cases (e.g., DDT, Hg) their partitioning between, or among, phases is not known. For this reason, and because there are transformations between the various phases, it is frequently necessary to sample in all phases concurrently.

The relative importance of various removal processes depends on such properties as the density and shape of particles, vapor pressure, chemical reactivity, solubility, and hygroscopicity. For particulate matter, these properties should be known as a function of size. A knowledge of these parameters will aid in recognizing and understanding physical or chemical changes that might occur in samples during the collection process. Finally, such information is essential to the modeling of atmospheric transport and removal processes.

B.  SAMPLE CONTAMINATION

Extreme care must be taken during the collection and handling of atmospheric gaseous, particulate, and precipitation samples, especially in remote areas (Duce *et al.,* 1974; Moyers *et al.,* 1972; Zoller *et al.,* 1974). All the substances of interest to this Workshop are present in concentrations that are so low that the slightest contamination could render a sample useless. In addition to employing instrumental safeguards to minimize contamination, personnel handling equipment and samples must be thoroughly familiar with precautions to be taken and must be aware of the many sources of contamination that can arise. Hands, hair, and clothing of personnel should be covered when handling these samples, and maximum use should be made of laminar-flow clean stations.

Long sampling intervals obviously have greater potential for contamination from local sources, such as the engine exhaust from a ship sampling platform. Use can be made of the fact that combustion processes produce extremely large concentrations of Aitken particles. Thus, a recording Aitken nucleus counter operating at sampling sites will provide information on short-term air-quality changes (i.e., the incursion of air parcels affected by local pollutant sources) as well as longer-term trends. The control of sampling systems by wind-direction and velocity sensors or by Aitken nucleus counters (Duce *et al.,* 1974) ensures that samples are collected only during desirable, or "clean," conditions. However, on occasion, filter samples should be collected during conditions that favor contamination, in order to identify characteristic qualities of the contaminants.

The possibility of contamination may also be reduced by placing electrical devices, combustion apparatus, and oil-lubricated machinery as far as possible from the collection apparatus. Materials or structures in contact with, or adjacent to, samplers should be chemically compatible with the analyses to be done, in order to avoid sample contamination.

## C. SAMPLING SITES

Features of various sampling platforms are listed in Table 10.2. Prior to establishing a permanent sampling site, it is essential that tests be carried out to ensure that the site is representative of the area being sampled and that it is free from contamination. Smoke bombs, released at upwind locations, may be useful to demonstrate that sea spray or local soil dust is not reaching the collecting location.

Oceanographic ships are usually not designed to accommodate the requirements of atmospheric chemistry sampling. In addition, the airflow around the ship structure is very complex. For such work, space, free from contamination, is required in the bow, high above the water.

Although gas and oil-well platforms may be used to collect atmospheric samples, the potential for contamination is also very great for the same reasons as it is aboard ships. Remote, small island stations can be useful. Currently, the United States is monitoring precipitation chemistry on the islands of Hawaii and American Samoa (Miller, 1975b).

Greater use will be made of aircraft as sampling platforms to gather information on the variations in concentration and properties of chemical species throughout the troposphere. Many techniques previously discussed will be adaptable for use in aircraft. Although aircraft may not be particularly useful for collecting precipitation samples (Table 10.2), they could be valuable for the collection of cloud-water samples. There have been few measurements of cloud-water chemistry as a precursor to precipitation, even at ground level in orographic clouds (Houghton, 1955; Martens and Harris, 1973). Currently, aircraft sampling of this type is being done by Petrenchuk (as referred to by Miller, 1975a). However, the relationship between the chemistry of cloud water and that of precipitation at ground or sea level is poorly understood.

## D. SUPPORTING INFORMATION

In any atmospheric measurement program, it is important that as much as possible be known about the past history of the air being sampled and about its state at the time of measurement. Meteorological measurements should include wind speed and direction and a time record of precipitation. Surface synoptic charts and 850-, 700-, and 500-mbar upper-air charts may be used to investigate the past history of the air parcels. Correlation of air-mass trajectories with pollutant concentrations will be a critical part of any meaningful program on atmospheric transport. Computer programs are now becoming available for determining air-parcel trajectories (see Chapter 3).

When collecting airborne particles, simultaneous measurements of several aerosol parameters, such as number, concentration, and size distribution (Hogan *et al.*, 1975), should be made in order to better characterize the entire aerosol.

### E.   DRY DEPOSITION MEASUREMENTS

The term "dry deposition" refers to material deposited on surfaces by all processes other than precipitation. This includes direct gas-phase transfer, impaction of particles, gravitational sedimentation of large particles, and diffusion processes for gases and small particles. The fluxes of certain inorganic gases to surfaces may be determined using micrometeorological techniques (flux-gradient or eddy correlation) under very restrictive conditions. Few such determinations have been attempted over the oceans.

It would be highly desirable to measure directly the chemical composition and rates of particulate deposition. However, measurements of this type are extremely difficult, and there has been only limited progress to date. Dustfall measurements, which have been made for many years, are at best semiquantitative; only the relatively small number of particles larger than about 10 $\mu$m are collected in this way. Similar reservations may be expressed about the use of the "dry" portion of the automatic "wet–dry" precipitation collectors now available to measure dry fallout. Such devices are subject to electrostatic effects and to aerodynamic effects induced by the collection device itself. Their collection and retention efficiencies are unknown and probably will vary with environmental conditions from site to site. Their value may lie in correcting bulk precipitation measurements for dry contamination.

A variety of techniques has been used to estimate the impaction of particles on natural surfaces; for example, with coniferous trees (Rosinski and Nagamoto, 1965; Langer, 1965), Spanish moss (Sheline *et al.*, 1976; Roberts, 1972), plastic coniferous trees (Schlesinger and Reiners, 1974), and filter paper (White and Turner, 1970; Peirson *et al.*, 1973, 1974). The last-referenced study collected samples of airborne dust, rainwater, and "dry deposition" concurrently to provide a useful indication of the relative "dry deposition" behavior of the various elements in the atmospheric aerosol.

However, data of this sort probably should be considered as provi-

TABLE 10.2 Types of Locations for Collection of Precipitation or Aerosols

| Platform | Advantages | Limitations | Problems of Contamination[a] | Research (R) or Monitoring (M) Use[b] | Precipitation (P) or Aerosol (A) Collections[b] |
|---|---|---|---|---|---|
| Continental coastline | Permanent, accessible, stable platform; electrical power readily available | Site restriction; sampling only during off-ocean flow; extrapolation to marine conditions required | Terrestrial dust, organic debris, birds, local anthropogenic emissions | R, M | P, A |
| Small island coastline | Permanent, stable platform; proximity to marine environment | Accessibility (?); restricted site choice; electrical power availability (?) | Terrestrial dust, organic debris, birds | R, M | P, A |
| Offshore structure (gas, oil well) | Permanent (?), stable platform | No choice of site; accessibility (?); electrical power (?) | Birds, local anthropogenic emissions, splash, etc. from superstructure | R, (M) | A, (P) |

| Platform | Advantages | Limitations | Interferences[a] | | |
|---|---|---|---|---|---|
| Ship | Mobility; access to analytical laboratory facilities and personnel on research vessels; electrical power | Unstable platform; mobility; restricted site selection and schedule (?) | Birds, local anthropogenic emissions; splash. etc. from superstructure; Samples should be taken from bow when under way and heading into wind | R, (M) | A, (P) |
| Buoy | Clean site; permanent (?) | Unstable platform; electrical power availability (?); height limitation | Birds, splash, etc. from superstructure | M, (R) | (P), (A) |
| Aircraft | Mobility; versatility | Air volumes (sampling time) limited; altitude and location restrictions; cost per unit time extremely high | Can be avoided | R | A |

[a] Site-specific sea spray (surf, ships wake, etc.) in all cases except aircraft.
[b] Items in parentheses indicate acceptable but not ideal conditions.

sional at this time over land surfaces, and they are nonexistent for the oceans.

## F.  VERTICAL DISTRIBUTION OF SUBSTANCES IN THE ATMOSPHERE

In order to assess the flux of materials to the oceans, it will be necessary to have information on the vertical distribution of concentrations as well as the areal distribution. These data must be obtained by aircraft. Because of the high cost of aircraft time, and because of operational limitations, we can never expect to develop an extensive data base for the atmosphere above the surface layer by this means. Eventually, remote-sensing techniques might be developed that would be suitable for this purpose, but these are still in the development stage. However, at present, there is available a relatively simple hand-held, inexpensive device—a sun photometer—that can provide information on the vertically integrated concentration of aerosols in the atmosphere, i.e., the atmospheric turbidity. The photometer measures the extinction of direct solar radiation (in one or more narrow passbands) due to scatter and absorption by aerosols that are primarily in the size range of 0.1 to 1.0 $\mu$m in diameter. Routine photometer measurements are made in a network of stations in the United States (Flowers et al., 1969) and in the World Meteorological Organization monitoring program (WMO, 1975). Large-scale patterns of turbidity have been related to major natural and anthropogenic sources of aerosols (Flowers et al., 1969; Volz, 1970). Measurements of extinction in several passbands are especially useful because they provide information about the aerosol size distribution (Roosen et al., 1973); on this basis, a distinction can be made between air masses bearing aerosols from arid regions and those coming from other continental areas (Roosen et al., 1973; Volz, 1970).

## IV.  RECOMMENDATIONS

1.  The collection of precipitation at sea remains a major problem. The difficulties are of two sorts: obtaining quantitative samples and obtaining samples free from contamination. A workshop should be convened on the subject of precipitation collection techniques and strategies for the marine environment.

2.  Over the past decade, there has been a great increase in the number of types of aerosol-collection instruments in use; these include

instruments that perform an aerosol-size classification function. In most cases, the sampling efficiency of these instruments under natural ambient conditions (or their laboratory equivalent) has not been determined. We recommend that a workshop be held on the subject of aerosol-collection techniques and that a field intercomparison be carried out in conjunction with the workshop or subsequent to it. Following the workshop and the field intercomparison, a set of recommendations should be compiled that specify preferred minimum design parameters for aerosol collectors.

3.   The routine use of sun photometers in aerosol programs is to be recommended. However, these instruments in their present form suffer from a number of shortcomings, including calibration drift and a lack of sensitivity for measurements in regions where aerosol concentrations are low. A strong effort should be made to improve instrument design.

## REFERENCES

ACGIH (1972). *Air Sampling Instruments,* 4th ed., American Conference of Governmental Industrial Hygienists, P.O. Box 1937, Cincinnati, Ohio.

Belyaev, S. P., and L. M. Levin (1974). Techniques for collection of representative aerosol samples, *J. Aerosol Sci. 5,* 325–338.

Bidleman, T. E., and C. E. Olney (1974). Chlorinated hydrocarbons in the Sargasso Sea atmosphere and surface water, *Science 183,* 516–518.

Butcher, S. S., and R. J. Charlson (1972). *An Introduction to Air Chemistry,* Academic Press, New York.

Dams, R., K. A. Rahn, and J. W. Winchester (1972). Evaluation of filter materials and impaction surfaces for non-destructive neutron activation analysis of aerosols, *Environ. Sci. Technol. 6,* 441–448.

Davies, C. N. (1968). The sampling of aerosols—the entry of aerosols into sampling tubes and heads, *Staub-Reinhalt Luft 28,* 1–9.

Delany, A. C., D. W. Parkin, J. J. Griffin, E. D. Goldberg, and B. E. F. Reimann (1967). Airborne dust collected at Barbados, *Geochim. Cosmochim. Acta 31,* 885–909.

Duce, R. A., G. L. Hoffman, J. L. Fasching, and J. L. Moyers (1974). The collection and analysis of trace elements in atmospheric particulate matter over the North Atlantic Ocean, *Special Environmental Report No. 3.* WMO, Geneva, pp. 370–379.

Duce, R. A., M. G. Gross, J. D. Hem, and K. K. Turekian (1976). Transport paths, *Strategies for Marine Pollution Monitoring,* E. D. Goldberg, ed., John Wiley & Sons, New York, Chap. 15.

Dzubay, T. G., L. E. Hines, and R. K. Stevens (1976). Particle bounce errors in cascade impactors, *Atmos. Environ. 10,* 229–234.

Elliott, W. P., and R. K. Reed (1973). Oceanic rainfall off the Pacific northwest coast, *J. Geophys. Res. 78,* 941–948.

Eriksson, E. (1959). The yearly circulation of chloride and sulphur in nature: meteorological, geochemical and pedological implications, Part I, *Tellus 11,* 375–403.

Eriksson, E. (1960). The yearly circulation of chloride and sulphur in nature: meteorological, geochemical and pedological implications, Part II, *Tellus 12,* 63–109.

Fasching, J. L., R. A. Courant, R. A. Duce, and S. R. Piotrowicz (1974). A new surface microlayer sampler utilizing the bubble microtome, *J. Rech. Atmos. 8*, 649–652.

Flowers, E. C., R. A. McCormick, and K. R. Kurtis (1969). Atmospheric turbidity over the United States, 1961–1966, *J. Appl. Meteorol. 8*, 955–962.

Forrest, J., and L. Newman (1973). Ambient air monitoring for sulphur compounds, a critical review, *J. Air Pollut. Control Assoc. 23*, 761–768.

Galloway, J. N., and G. E. Likens (1975). Calibration of collection procedures for determination of precipitation chemistry, *Proc. First International Symposium on Acid Precipitation and the Forest Ecosystem*, USDA Forest Service General Tech. Rep. NE-23, Upper Darby, Pa., pp. 137–156.

Gambell, A. W., and D. W. Fisher (1966). Chemical composition of rainfall, eastern North Carolina and southeastern Virginia, U.S. Geological Survey Water-Supply Paper 1535-K, 44 pp.

Grimsrud, E. P., and R. A. Rasmussen (1975). The analysis of chlorofluorocarbons in the troposphere by gas chromatography–mass spectrometry, *Atmos. Environ. 9*, 1010–1013.

Harvey, G. R., and W. G. Steinhauer (1974). Atmospheric transport of polychlorobiphenyls to the North Atlantic, *Atmos. Environ. 8*, 777–782.

Hogan, A., W. Winters, and G. Gardnew (1975). A portable aerosol detector of high sensitivity, *J. Appl. Meteorol. 14*, 39–45.

Houghton, H. G. (1955). On the chemical composition of fog and cloud water, *J. Meteorol. 12*, 355–357.

Hounam, R. F., and R. J. Sherwood (1965). The cascade centripeter: a device for determining the concentration and size distribution of aerosols, *Am. Ind. J. 26*, 122–131.

Johnson, D. A., and D. H. F. Atkins (1974). A technique for measurement of sulphur dioxide and sulphate concentrations from an aircraft, *Special Environmental Report No. 3*, WMO, Geneva, pp. 216–222.

Junge, C. E. (1963). *Air Chemistry and Radioactivity*, Academic Press, New York, 382 pp.

Jutze, G. A., and K. E. Foster (1967). Recommended standard method for atmospheric sampling of fine particulate matter by filter media-high volume sampler, *J. Air Pollut. Control Assoc. 17*, 17–25.

Langer, G. (1965). Particle deposition and re-entrainment from coniferous trees, Part II: Experiments with individual leaves, *Kolloid Z. Z. Polym. 204*, 119–124.

Likens, G. E., F. H. Bormann, N. M Johnson, and R. S. Pierce (1967). The calcium, magnesium, potassium and sodium budgets for a small forested ecosystem, *Ecology 48*, 772–785.

Liu, B. Y. H., and K. W. Lee (1976). Efficiency of membrane and Nuclepore filters for submicrometer aerosols, *Environ. Sci. Technol. 10*, 345–350.

Lodge, J. P. Jr., K. C. Hill, J. B. Pate, E. Lorange, W. Basbergill, A. L. Lazrus, and G. S. Swanson (1968). Chemistry of the United States precipitation, Final report on the national precipitation sampling network, Laboratory of Atmosphere Sciences, National Center for Atmospheric Research, Boulder, Colo., 66 pp.

Lovelock, J. E., R. J. Maggs, and R. J. Wade (1973). Halogenated hydrocarbons in and over the Atlantic, *Nature 241*, 194–196.

Marple, V. A., and B. Y. H. Liu (1974). Characteristics of laminar jet impactors, *Environ. Sci. Technol. 8*, 648–654.

Marple, V. A., and B. Y. H. Liu (1975). On fluid flow and aerosol impaction in inertial impactors, *J. Colloid Interface Sci. 53*, 31–34.

Marten, C. S., and R. C. Harris (1973). Chemistry of aerosols, cloud droplets and rain in the Puerto Rican marine atmosphere, *J. Geophys. Res. 76*, 949–957.

May, K. R., N. P. Pomeroy, and S. Hibbs (1976). Sampling techniques for large windborne particles, *J. Aerosol Sci. 7*, 53–62.

Miller, J. M. (1975a). Visit to the Soviet Union concerning precipitation chemistry and other geophysical monitoring programs. Trip report to Nat. Oceanic and Atmospheric Administration, U.S. Dept. of Commerce, Washington, D.C.

Miller, J. M., ed. (1975b). *Geophysical Monitoring for Climatic Change*, No. 3, Summary Rep. 1974, U.S. Dept. of Commerce, NOAA, 100 pp.

Moyers, J. L., R. A. Duce, and G. L. Hoffman (1972). A note on the contamination of atmospheric particulate samples collected from ships, *Atmos. Environ. 6*, 551–556.

Natusch, D. F. S., and J. R. Wallace (1976). Determination of airborne particle size distributions and calculation of cross-sensitivity and discreteness effects in cascade impaction, *Atmos. Environ. 10*, 315–324.

Paterson, M. P. (1975). The atmospheric transport of natural and man-made substances, Ph.D. Thesis, U. of London, 132 pp.

Peirson, D. H., P. A. Cawse, L. Salmon, and R. S. Cambray (1973). Trace elements in the atmospheric environment, *Nature 241*, 252–256.

Peirson, D. H., P. A. Cawse, and R. S. Cambray (1974). Chemical uniformity of airborne particulate material, and a maritime effect, *Nature 251*, 675–679.

Petrenchuk, O. P., and E. S. Selezneva (1970). Chemical composition of precipitation in region of the Soviet Union, *J. Geophys. Res. 75*, 3629–3634.

Prospero, J. M., and R. T. Nees (1977). Dust concentration in the atmosphere of the equatorial North Atlantic: Possible relationship to the Sahelian drought, *Science 196*, 1196–1198.

Roberts, R. M. (1972). Plants as monitors of airborne metal pollution, *J. Environ. Planning Pollut. Control 1*, 3–5.

Roosen, R. G., R. J. Angione, and C. H. Klemcke (1973). Worldwide variations in atmospheric transmission: 1. Baseline results from Smithsonian observations, *Bull. Am. Meteorol. Soc. 54*, 307–316.

Rosinski, J., and C. T. Nagamoto (1965). Particle deposition and re-entrainment from coniferous trees. Part I: Experiments with trees, *Kolloid Z. Z. Polym. 204*, 111–119.

Schlesinger, W. H., and W. A. Reiners (1974). Deposition of water and cations on artificial foliar collectors in fir krummholz of New England mountains, *Ecology 55*, 378–386.

Selezneva, E. S. (1974a). The investigation of the transport of sea salt from the Black Sea coast by using precipitation chemistry data, *Trudy MGO 343*, 34–45 (in Russian).

Selezneva, E. S. (1974b). Some physical chemical characteristics of atmospheric precipitation at the coast of the Pacific, *Trudy MGO 343*, 46–53 (in Russian).

Sheline, J., R. Akselsson, and J. W. Winchester (1976). Trace element similarity groups in North Florida Spanish moss: evidence for direct uptake of aerosol particles, *J. Geophys. Res. 81*, 1047–1050.

Simpson, J. H. (1972). Aerosol cations at Mauna Loa Observatory, *J. Geophys. Res. 77*, 5266–5277.

Smith, D. K. (1975). Sampling and measurement networks: rainfall and dustfall sampler evaluations, *J. Great Lakes Res. 2, Suppl. 1*, 78–81.

Spurney, K. R., J. P. Lodge, E. R. Frank, and D. C. Sheesley (1969). Aerosol filtration by means of Nuclepore filters: structural and filtration properties, *Environ. Sci. Technol. 3*, 453.

Stafford, R. G., and J. H. Ettinger (1972). Filter efficiency as a function of particle size and velocity, *Atmos. Environ. 6*, 353–362.

Stevens, R. K., and T. G. Dzubay (1975). Recent developments in air particulate monitoring. *Proc. Instr. Methods for Air and Water Measurements and Monitoring, IEEE Trans. Nucl. Sci.*, 18 pp.

Su, C.-W., and E. D. Goldberg (1973). Chlorofluorocarbons in the atmosphere, *Nature* *245*, 27.

Summers, P. W., and D. M. Whelpdale (1975). Acid precipitation in Canada, *Proc. First International Symposium on Acid Precipitation and the Forest Ecosystem*, Columbus, Ohio.

Volz, F. (1970). Spectral skylight and solar radiance measurements in the Caribbean: Maritime aerosols and Sahara dust, *J. Atmos. Sci. 27*, 1041–1047.

Wedding, J. B., A. R. McFarland, and J. E. Cermak (1977). Large particle collection characteristics of ambient aerosol samplers, *Environ. Sci. Technol. 11*, 387–390.

White, E. J., and F. Turner (1970). A method of estimating income of nutrients in catch of airborne particles by a woodland canopy, *J. Appl. Ecol. 7*, 441–461.

Windom, H. L., and R. A. Duce (1976). *Marine Pollutant Transfer*, Lexington Books, (D. C. Heath), Lexington, Mass.

WMO (1975). *WMO Operations Manual for Sampling and Analysis Techniques for Chemical Constituents in Air and Precipitation*, WMO, Geneva.

Zoller, W. H., E. S. Gladney, R. A. Duce, and G. L. Hoffman (1974). Trace metals in the Antarctic atmosphere, *Special Environmental Report No. 3.*, WMO, Geneva, pp. 380–384.

# Appendix: The United Nations Directory of Existing Pollution Monitoring Programs

R. CITRON

The United Nations Environment Program (UNEP) has a mandate to develop a Global Environmental Monitoring System (GEMS) as part of its Earthwatch Program. One of the first priorities of GEMS is to coordinate and support existing pollution monitoring programs that provide data and information that may be used to assess the current states and trends of global pollution. UNEP has developed a computerized directory on the world's existing pollution monitoring programs. The directory now includes information on the operational and administrative characteristics of 700 pollution-monitoring programs operating more than 66,000 permanent monitoring and sampling sites in 78 countries on every continent and ocean of the world. Many of these programs are already systematically sampling pollutants that are transported by the atmosphere, and the directory can be used to design *ad hoc* networks to sample specific pollutants in specific media (air, precipitation, marine water, etc.) at, or near, their sources, to track their transport and fallout en route to, and in, the ocean, and to measure levels at remote sites to provide background data.

As an example, there are at least 42 separate programs in 28 countries that utilize over 3000 sites to sample air, precipitation, and marine water for chlorinated hydrocarbons, toxic heavy metals, petroleum hydrocarbons, selected radionuclides, and selected trace gases. The directory can be used to design pollution sampling networks utilizing existing facilities by selecting sites for geographical location, means of sample acquisition, sampling frequency, measurement or

241

analysis technique, mode of data storage, or any combination of these parameters. The directory can also be used to identify gaps in existing networks for any selected pollutant in any specified medium. While the directory does not contain information on all pollution monitoring programs in all countries that have them, it does represent a first approximation of the world's existing pollution monitoring resources.

## A DESCRIPTION OF THE UNITED NATIONS POLLUTION MONITORING PROGRAM DIRECTORY

The computerized directory is designed to be used as a major resource for the planning of integrated networks of pollution monitoring sites that would systematically measure pollutant levels throughout the world. It contains information on the purpose of each program, how the programs are administered, and on the pollutants they monitor. The following information on the administrative characteristics of each program is contained in the directory:

1. Program name;
2. Purpose of program;
3. Name and address of administrative organization or office and the office responsible for storage of data collected in the program;
4. Director's name, address, telephone and telex numbers, and cable address;
5. List of priority pollutants monitored.

For each of the priority pollutants, the following operational or technical information is contained in the data base:

1. Physical medium monitored;
2. Geographical area covered;
3. Number of monitoring sites;
4. Sampling frequency, measurement techniques, and precision of analysis;
5. Means of data or sample acquisition;
6. Mode of data storage (publications, punch cards, magnetic tape, etc.);
7. Latitude and longitude of each monitoring site in each program, when available.

The directory can supply answers to specific questions about the data. One may request a list of programs that are monitoring a specified pollutant in a given medium in a certain group of countries. The data base also can supply statistics on the number of programs sharing the same monitoring characteristics such as monitoring methodology, sampling frequency, medium monitored, and analysis techniques. Furthermore, a computerized plotting machine can draw world or regional maps indicating the location of sites selected according to common characteristics. Plat maps and computer printouts will greatly simplify the design and implementation of planned pollutant-monitoring networks.

X